188
Advances in Polymer Science

Editorial Board:
A. Abe · A.-C. Albertsson · R. Duncan · K. Dušek · W. H. de Jeu
J.-F. Joanny · H.-H. Kausch · S. Kobayashi · K.-S. Lee · L. Leibler
T. E. Long · I. Manners · M. Möller · O. Nuyken · E. M. Terentjev
B. Voit · G. Wegner · U. Wiesner

Advances in Polymer Science

Recently Published and Forthcoming Volumes

Conformation-Dependent Design of Sequences in Copolymers II
Volume Editor: Khokhlov, A. R.
Vol. 196, 2006

Conformation-Dependent Design of Sequences in Copolymers I
Volume Editor: Khokhlov, A. R.
Vol. 195, 2006

Enzyme-Catalyzed Synthesis of Polymers
Volume Editors: Kobayashi, S., Ritter, H., Kaplan, D.
Vol. 194, 2006

Polymer Therapeutics II
Polymers as Drugs, Conjugates and Gene Delivery Systems
Volume Editors: Satchi-Fainaro, R., Duncan, R.
Vol. 193, 2006

Polymer Therapeutics I
Polymers as Drugs, Conjugates and Gene Delivery Systems
Volume Editors: Satchi-Fainaro, R., Duncan, R.
Vol. 192, 2006

Interphases and Mesophases in Polymer Crystallization III
Volume Editor: Allegra, G.
Vol. 191, 2005

Block Copolymers II
Volume Editor: Abetz, V.
Vol. 190, 2005

Block Copolymers I
Volume Editor: Abetz, V.
Vol. 189, 2005

Intrinsic Molecular Mobility and Toughness of Polymers II
Volume Editor: Kausch, H.-H.
Vol. 188, 2005

Intrinsic Molecular Mobility and Toughness of Polymers I
Volume Editor: Kausch, H.-H.
Vol. 187, 2005

Polysaccharides I
Structure, Characterization and Use
Volume Editor: Heinze, T.
Vol. 186, 2005

Advanced Computer Simulation Approaches for Soft Matter Sciences II
Volume Editors: Holm, C., Kremer, K.
Vol. 185, 2005

Crosslinking in Materials Science
Vol. 184, 2005

Phase Behavior of Polymer Blends
Volume Editor: Freed, K.
Vol. 183, 2005

Polymer Analysis/Polymer Theory
Vol. 182, 2005

Interphases and Mesophases in Polymer Crystallization II
Volume Editor: Allegra, G.
Vol. 181, 2005

Interphases and Mesophases in Polymer Crystallization I
Volume Editor: Allegra, G.
Vol. 180, 2005

Inorganic Polymeric Nanocomposites and Membranes
Vol. 179, 2005

Polymeric and Inorganic Fibres
Vol. 178, 2005

Intrinsic Molecular Mobility and Toughness of Polymers II

Volume Editor: Hans-Henning Kausch

With contributions by
V. Altstädt · M. C. Baietto-Dubourg · C.-M. Chan · A. Chateauminois
R. Estevez · E. Van der Giessen · C. Grein · L. Li

Springer

The series *Advances in Polymer Science* presents critical reviews of the present and future trends in polymer and biopolymer science including chemistry, physical chemistry, physics and material science. It is adressed to all scientists at universities and in industry who wish to keep abreast of advances in the topics covered.

As a rule, contributions are specially commissioned. The editors and publishers will, however, always be pleased to receive suggestions and supplementary information. Papers are accepted for *Advances in Polymer Science* in English.

In references *Advances in Polymer Science* is abbreviated *Adv Polym Sci* and is cited as a journal.

Springer WWW home page: http://www.springeronline.com
Visit the APS content at http://www.springerlink.com/

Library of Congress Control Number: 2005926289

ISSN 0065-3195
ISBN-10 3-540-26162-1 Springer Berlin Heidelberg New York
ISBN-13 978-3-540-26162-9 Springer Berlin Heidelberg New York
DOI 10.1007/b136969

This work is subject to copyright. All rights are reserved, whether the whole or part of the material is concerned, specifically the rights of translation, reprinting, reuse of illustrations, recitation, broadcasting, reproduction on microfilm or in any other way, and storage in data banks. Duplication of this publication or parts thereof is permitted only under the provisions of the German Copyright Law of September 9, 1965, in its current version, and permission for use must always be obtained from Springer. Violations are liable for prosecution under the German Copyright Law.

Springer is a part of Springer Science+Business Media

springeronline.com

© Springer-Verlag Berlin Heidelberg 2005
Printed in Germany

The use of registered names, trademarks, etc. in this publication does not imply, even in the absence of a specific statement, that such names are exempt from the relevant protective laws and regulations and therefore free for general use.

Cover design: *Design & Production* GmbH, Heidelberg
Typesetting and Production: LE-TEX Jelonek, Schmidt & Vöckler GbR, Leipzig

Printed on acid-free paper   02/3141 YL – 5 4 3 2 1 0

Volume Editor

Prof. Dr. Hans-Henning Kausch
Ecole Polytechnique Fédérale de Lausanne
Science de Base
Station 6
1015 Lausanne, Switzerland
kausch.cully@bluewin.ch

Editorial Board

Prof. Akihiro Abe
Department of Industrial Chemistry
Tokyo Institute of Polytechnics
1583 Iiyama, Atsugi-shi 243-02, Japan
aabe@chem.t-kougei.ac.jp

Prof. A.-C. Albertsson
Department of Polymer Technology
The Royal Institute of Technology
10044 Stockholm, Sweden
aila@polymer.kth.se

Prof. Ruth Duncan
Welsh School of Pharmacy
Cardiff University
Redwood Building
King Edward VII Avenue
Cardiff CF 10 3XF
United Kingdom
duncan@cf.ac.uk

Prof. Karel Dušek
Institute of Macromolecular Chemistry,
Czech
Academy of Sciences of the Czech Republic
Heyrovský Sq. 2
16206 Prague 6, Czech Republic
dusek@imc.cas.cz

Prof. Dr. W. H. de Jeu
FOM-Institute AMOLF
Kruislaan 407
1098 SJ Amsterdam, The Netherlands
dejeu@amolf.nl
and Dutch Polymer Institute
Eindhoven University of Technology
PO Box 513
5600 MB Eindhoven, The Netherlands

Prof. Jean-François Joanny
Physicochimie Curie
Institut Curie section recherche
26 rue d'Ulm
75248 Paris cedex 05, France
jean-francois.joanny@curie.fr

Prof. Dr. Hans-Henning Kausch
Ecole Polytechnique Fédérale de Lausanne
Science de Base
Station 6
1015 Lausanne, Switzerland
kausch.cully@bluewin.ch

Prof. S. Kobayashi
R & D Center for Bio-based Materials
Kyoto Institute of Technology
Matsugasaki, Sakyo-ku
Kyoto 606-8585, Japan
kobayash@kit.ac.jp

Prof. Kwang-Sup Lee
Department of Polymer Science &
Engineering
Hannam University
133 Ojung-Dong Taejon
300-791, Korea
kslee@mail.hannam.ac.krr

Prof. L. Leibler
Matière Molle et Chimie
Ecole Supérieure de Physique
et Chimie Industrielles (ESPCI)
10 rue Vauquelin
75231 Paris Cedex 05, France
ludwik.leibler@espci.fr

Prof. Timothy E. Long
Department of Chemistry
and Research Institute
Virginia Tech
2110 Hahn Hall (0344)
Blacksburg, VA 24061, USA
telong@vt.edu

Prof. Ian Manners
School of Chemistry
University of Bristol
Cantock's Close
BS8 1TS Bristol, UK
r.musgrave@bristol.ac.uk

Prof. Dr. Martin Möller
Deutsches Wollforschungsinstitut
an der RWTH Aachen e.V.
Pauwelsstraße 8
52056 Aachen, Germany
moeller@dwi.rwth-aachen.de

Prof. Oskar Nuyken
Lehrstuhl für Makromolekulare Stoffe
TU München
Lichtenbergstr. 4
85747 Garching, Germany
oskar.nuyken@ch.tum.de

Dr. E. M. Terentjev
Cavendish Laboratory
Madingley Road
Cambridge CB 3 OHE
United Kingdom
emt1000@cam.ac.uk

Prof. Brigitte Voit
Institut für Polymerforschung Dresden
Hohe Straße 6
01069 Dresden, Germany
voit@ipfdd.de

Prof. Gerhard Wegner
Max-Planck-Institut
für Polymerforschung
Ackermannweg 10
Postfach 3148
55128 Mainz, Germany
wegner@mpip-mainz.mpg.de

Prof. Ulrich Wiesner
Materials Science & Engineering
Cornell University
329 Bard Hall
Ithaca, NY 14853
USA
ubw1@cornell.edu

Advances in Polymer Science
Also Available Electronically

For all customers who have a standing order to Advances in Polymer Science, we offer the electronic version via SpringerLink free of charge. Please contact your librarian who can receive a password or free access to the full articles by registering at:

springerlink.com

If you do not have a subscription, you can still view the tables of contents of the volumes and the abstract of each article by going to the SpringerLink Homepage, clicking on "Browse by Online Libraries", then "Chemical Sciences", and finally choose Advances in Polymer Science.

You will find information about the

- Editorial Board
- Aims and Scope
- Instructions for Authors
- Sample Contribution

at springeronline.com using the search function.

Preface

The enormous length of macromolecules and the low intra- and intermolecular barriers opposing rotation and displacement of molecular groups or of even longer segments are at the origin of the unique visco- and rubber-elastic behaviour of polymer solids. Molecular mobility influences all phases of processing and use of such materials. Thus segregation and phase separation in the melt as well as structure development through crystallization depend on chain dynamics. The same is true for most deformation mechanisms, sample stiffness and ultimate properties such as toughness. Considerable progress has been obtained in the last decade in the understanding of the mutual relationship between the primary molecular parameters chain configuration, architecture and molecular weight (MW) on the one hand, and the response of a loaded entanglement network, the nature of the processes limiting stress transfer and the resulting mode of mechanical breakdown on the other. In view of the large technical importance of mechanical performance it seems to be adequate to review this subject, the *Intrinsic Molecular Mobility and Toughness of Polymers*.

In their introductory contribution Kausch and Michler discuss the elementary, time-dependent molecular deformation mechanisms, the competition between them, and their influence on the different failure modes of thermoplastic polymers (crazing, creep, yielding and flow, fracture through crack propagation). By establishing a *micro-morphological model* of polymer deformation and durability the authors highlight the dual role of segmental jumps and displacements to improve toughness by energy dissipation and relaxation of critical stresses and to influence without exception all damage mechanisms.

The dynamic response of a chain segment to thermo-mechanical excitation strongly depends on in-chain cooperative motions. By combining the powerful techniques of multi-dimensional Nuclear Magnetic Resonance and of dielectric and dynamic mechanical analysis Monnerie, Laupêtre and Halary have investigated the *intensity and molecular origin of sub-T_g relaxations* and their degree of coupling for five structurally quite different amorphous polymers. Their important findings are reported in two comprehensive reviews treating the effect of chain configuration on segmental mobility and its effect on the toughness of these materials, respectively.

Essential features of the entanglement network and of the morphology of semi-crystalline polymers are determined through the crystallization process.

Chan and Li review homogeneous and heterogeneous nucleation. Using the new hot-stage in-situ AFM technique they particularly investigate the propagation of *founding lamellae*, their branching, interaction and development into lamellar sheaves and spherulites. In her contribution Grein gives a thorough *analysis of the influence of phase structure* (α- and β-crystalline polypropylene) as compared to the effect of elastomeric modifier particles. She concludes that the capacity of a matrix to deform remains an essential requirement for high toughness materials.

Stress cracking environments are known to enhance the mobility in the affected surface regions. Altstädt shows that the rate of fatigue crack propagation at *constant stress intensity factor K* proves to be a sensitive quantitative measure of the influence of active media. He also points to the dual role of segmental mobility, permitting stress relaxation followed by strain hardening or unstable softening, respectively. The complex conditions of *fracture during sliding contact* are reviewed by Chateauminois and Baietto-Duboug. They arrive at the conclusion that the main wear mechanism of glassy polymers, asperity scratching, is strongly controlled by competition between crazing processes and shear yielding. In the final contribution Estevez and van der Giessen present a computational analysis of the fracture of glassy polymers. The *applied cohesive zone model* takes into consideration the three steps of crazing (initiation, thickening and breakdown) and seems to be sufficiently flexible to adapt to future refinements.

The editor wishes to thank all authors for their willingness to cooperate in this joint effort, which so heavily depended on the concourse of their special expertise. It is hoped that the resulting detailed overview will be of help to more fully exploit the large potential offered by polymeric systems. Unfortunately the comprehensive treatment has made it necessary to publish the above, closely related eight contributions in two consecutive volumes of the Advances in Polymer Science, Vols. 187 and 188. However, a common *Subject Index* in both volumes and the reproduction of the two *List of Contents* should make it easy for the reader to find the desired information.

Lausanne, September 2005 *Hans-Henning Kausch*

Contents

Direct Observation of the Growth of Lamellae and Spherulites by AFM
C.-M. Chan · L. Li 1

Toughness of Neat, Rubber Modified and Filled β-Nucleated Polypropylene:
From Fundamentals to Applications
C. Grein 43

The Influence of Molecular Variables on Fatigue Resistance in Stress Cracking
Environments
V. Altstädt 105

Fracture of Glassy Polymers Within Sliding Contacts
A. Chateauminois · M. C. Baietto-Dubourg 153

Modeling and Computational Analysis of Fracture of Glassy Polymers
R. Estevez · E. Van der Giessen 195

Author Index Volumes 101–188 235

Subject Index 259

Contents of Volume 187

Intrinsic Molecular Mobility and Toughness of Polymers I

Volume Editor: Hans-Henning Kausch
ISBN: 3-540-26155-9

The Effect of Time on Crazing and Fracture
H.-H. Kausch · G. H. Michler

Investigation of Solid-State Transitions in Linear and Crosslinked Amorphous Polymers
L. Monnerie · F. Lauprêtre · J. L. Halary

Deformation, Yield and Fracture of Amorphous Polymers: Relation to the Secondary Transitions
L. Monnerie · J. L. Halary · H.-H. Kausch

Direct Observation of the Growth of Lamellae and Spherulites by AFM

Chi-Ming Chan[1] (✉) · Lin Li[2]

[1] Department of Chemical Engineering, Hong Kong University of Science and Technology, Clear Water Bay, Hong Kong, P.R. China
kecmchan@ust.hk

[2] State Key Laboratory of Polymer Physics and Chemistry, Institute of Chemistry, Chinese Academy of Sciences, Peking, P.R. China
lilin@iccas.ac.cn

1	Introduction	2
2	Crystallization Processes	6
2.1	Homogeneous Nucleation—Birth of Primary Nuclei	6
2.2	Development of the Founding Lamella	9
3	Development of the Lamellar Sheaf	14
3.1	Branching of Lamellae	14
3.2	Branching at Different Temperatures	19
3.3	Propagation of Lamellae	22
3.4	Lamellar Growth Rate	25
4	The Effect of Film Thickness on Lamellar Growth Rate and Morphology	29
5	Formation of Spherulites	35
6	Heterogeneous Nucleation	37
7	Summary	39
	References	39

Abstract This article describes some of the progress made in the understanding of the growth of polymer lamellae and spherulites using atomic force microscopy (AFM) in the last five years. High-resolution and real-time AFM phase imaging enables us to observe the detailed growth process of lamellae. During the early stage of crystallization, embryos appear and disappear on the film surface. A stable embryo develops into a single lamella, which develops into a founding lamella. Then, the founding lamella develops into a lamellar sheaf through branching and splaying. Through further branching and splaying, a lamellar sheaf develops into a spherulite with two eyes at its center.

Keywords Atomic force microscopy · Branching · Crystallization · Embryo · Lamella · Polymer · Spherulite

Abbreviations

AFM	atomic force microscopy
EM	electron microscopy
T_g	glass transition temperature
i-PP	isotactic polypropylene
i-PS	isotactic polystyrene
OM	optical microscopy
BA-C10	poly(bisphenol A-co-decane)
BA-C8	poly(bisphenol A-co-octane)
PEO	poly(ethylene oxide)
PS[(S)-LA]	poly[(s)-lactide]
PCL	polycaprolactone
PE	polyethylene
PP	polypropylene
SEM	scanning electron microscopy
TM-AFM	tapping-mode atomic force microscopy
TEM	transmission electron microscopy
ΔG_{edge}	free energy of formation for an edge-on primary nucleus
ΔG_b	free energy change per unit of crystalline material formed
γ_f	interfacial energy between the folding surface and the melt
γ_ℓ	interfacial energy between the lateral plane and the melt
γ_{cs}	interfacial energy between the crystal and the substrate
γ_{ms}	interfacial energy between the melt and the substrate
a	dimension of a nucleus;
a_c	critical dimension of a nucleus
ℓ	dimension of a nucleus
ℓ_c	critical dimension of a nucleus
ΔG_{flat}	free energy of formation for a flat-on nucleus
ΔH_b	enthalpy of melting per unit volume of crystal
T_m^o	equilibrium melting point of a polymer
T_c	crystallization temperature
$\overline{M}_W$	weight-average molecular weight
g	average growth rate of a lamella
L_0	length between the lamellar tip and the location at which an induced nucleus just appears
t_i	induction time for the formation of an induced nucleus

1
Introduction

When a polymer crystallizes from the melt without disturbance, it normally forms spherical structures that are called spherulites [1, 2]. The dimensions of spherulites range from micrometers to millimeters, depending on the structure of the polymer chain and the crystallization conditions, such as cooling rate, crystallization temperature, and the content of the nucleating agent. The structure of spherulites is similar regardless of their size; they are aggregates of crystallites [1–6].

Much effort has been devoted to investigating the detailed architectures and the construction of spherulites. Early investigations of the crystallization of polymers through optical microscopy (OM) [7, 8] posited that polymer spherulites consisted of radiating fibrous crystals with dense branches to fill space. Later, when electron microscopy (EM) became available, spherulites were shown to be comprised of layer-like crystallites [9, 10], which were named lamellae. The lamellae are separated by disordered materials. In the center of the spherulites, the lamellae are stacked almost in parallel [5, 6, 11–15]. Away from the center, the stacked lamellae splay apart and branch, forming a sheaf-like structure [11, 13–15]. It was also found that the thicknesses of lamellae are different [5, 6, 11, 12]. The thicker ones are believed to be dominant lamellae while the thinner ones are subsidiary lamellae.

EM and OM have a few drawbacks, making them unsuitable for application in real-time studies of spherulitic growth at the lamellar level. The resolution of OM is too low to detect lamellae. The required sample preparation techniques for scanning electron microscopy (SEM) and transmission electron microscopy (TEM) usually stop the crystal growth. Examining the internal structure of growing spherulites in a partially crystallized and quenched polymer sample overcomes this drawback to a certain extent, and the process of spherulite formation has been deduced [4, 6, 11]. A spherulite is believed to develop from a stack of lamellae. During the growth process, the stacked lamellae splay apart continually and branch occasionally. The continuous growth of the dominant lamellae leads to the formation of a spherical skeleton and the subsidiary lamellae fill up the space between the dominant lamellae.

Phillips and Edwards studied the spherulitic morphology and the growth kinetics of natural, isomerized, and synthetic *cis*-polyisoprene under different pressures with TEM [16–23]. The polymer thin films were stained with osmium tetroxide vapor to stop the crystallization and enhance the phase contrast. Heterogeneous nucleation and homogeneous nucleation were both observed [18]. Heterogeneous nucleation was identified by the observations of lamellae growing in all directions normal to the surface of the nucleus. In homogeneous nucleation, lamellae were observed to grow in two directions. The growth rates of dominant and subsidiary lamellae were found to be the same and constant [19]. The spherulitic morphologies of melt-crystallized poly(4-methyl pentane) [24], polyethylene (PE) [10, 11, 13, 25–27], isotactic polypropylene (*i*-PP) [11], and isotactic polystyrene (*i*-PS) [28] were investigated using OM and TEM. The main points made by the authors concerning the spherulitic growth were that dominant lamellae first grew into the melt to form a skeleton of a spherulite by splaying and branching; inter-dominant lamellar regions were filled with subsidiary lamellae; and branching was mainly through giant screw dislocations. The pressure build-up by molecular cilia between lamellae caused the splaying.

Splaying apart and branching of lamellae to form spherulites due to the repulsion of the amorphous materials between the lamellae are the general features of polymer spherulites [4, 11, 14, 15]. It is understood that to achieve a spherical shape, primary lamellae have to splay apart and branch. Even so, the origins for the splaying and branching processes have not been confirmed. The early investigations of polymer crystallization suggested that the accumulation of noncrystallizable impurities in the front of growing lamellae was the reason for branching [29]. But several authors have queried this diffusion-control mechanism. On the basis of the morphology of growing spherulites in quenched samples, Bassett and his colleagues proposed that branching is a result of secondary nucleation on primary lamellae [11]. They also suggested that the mutual repulsion between the adjacent primary lamellae, resulting from the protruding cilia of the lamellae, is the origin for splaying of the stacked lamellae. However, the secondary nucleation and splaying processes have never been observed directly owing to the limitations of EM.

This process described, as shown schematically in Fig. 1, is still hypothetical because the description is not based on direct observations of the formation of spherulites. Furthermore, how the stacked lamellae are generated is still not clear, though nucleation of a supercooled melt can be predicted from thermodynamics [4]. The invention of atomic force microscopy (AFM) [30–36] has made direct observation of the crystallization of polymers possible. In addition to its high resolution, contact-mode AFM can record real-time images of a dynamic process. The disadvantage of contact-mode AFM is the considerable pressure exerted by the probe tip on the sample surface, causing sample deformation and induced nucleation. In particular, such damage to soft samples such as polymers and biological specimens limits the applicability of AFM. In recent years, the development of tapping-mode AFM (TM-AFM) has enhanced the capability of AFM as a surface analysis technique [35, 36]. In TM-AFM, a fast oscillating probe is used for surface imaging. During the operation, the tip makes contact with the surface briefly in each cycle of oscillation. Many studies have been performed to interpret the height and phase images recorded by TM-AFM [37–40]. The results clearly indicate that phase images can provide enhanced contrast on heterogeneous surfaces. TM-AFM has been shown to be a powerful tool to study the surfaces of polymer blends, copolymers, and semi-crystalline polymers [36–38, 41–44].

Fig. 1 Schematic showing the formation of a spherulite from a lamella

By utilizing the advantages of AFM, the spherulites and lamellae of various semi-crystalline polymers have been investigated [45–84]. The lamellar thickness [49, 68] and hedritic morphology [54] of polypropylene (PP) were studied by Vancso and colleagues and the thickness of the lamellae was found to be identical to the values in the literature as revealed by other techniques. The growth rates of a spherulite and the internal lamellae of a poly(hydroxybutyrate-co-valerate) copolymer have been determined by Hobbs and colleagues with TM-AFM [58]. The result indicated that the overall gross growth front of a spherulite can propagate at a constant rate although internal lamellae cannot propagate in the same way. It was posited that the spherulite growth rate is dependent on the rate of secondary nucleation on the existing lamellae but not on the growth rate of the lamellae. Melting and crystallization of poly(ethylene oxide) (PEO) [47, 57], PEO in PEO/poly(methyl methacrylate) blends [48], poly(ether ether ketone) [53], and PE [55] have been studied by an AFM equipped with a hot stage.

In order to record the dynamic polymer crystal growth process in-situ, two factors are significantly important. One is the use of a very high resolution technique. Such a technique can repeatedly record the same area without damage to or significant interactions with the sample. AFM has been proved to be a successful tool to fulfill this task. The other key factor is that the polymers must have an appropriate crystallization rate. It generally takes an AFM several minutes to produce an image. This requires that the polymer has a very slow crystallization rate. The crystallization rates of most semi-crystalline polymers at room temperature are too fast.

It is well known that the crystallization rate and crystallinity of a polymer are strongly dependent on the crystallization temperature and polymer chain structure. A polymer can be controlled to have a slow crystallization rate by crystallizing at a high temperature close to its melting point or a lower temperature near its glass transition temperature. It is much more difficult to obtain high-quality AFM images of polymers at high temperatures because the melt may interact with the AFM tip which may induce crystallization as well. To run AFM at low temperatures, a cooler is required to keep the sample cool. As a result, the operation at or near room temperature is preferred. Modifying the polymer chain structure can reduce the crystallization rate. Varying the flexible segment length can control the flexibility of the polymer chain, which affects its crystallization rate significantly. Li et al. prepared a series of polymers (BA-Cn) by phase-transfer catalyzed polyetherification of 1, n-dibromoalkane (Cn, n = 4, 6, 8, 10, 12, 14, and 18) with bisphenol A (BA), with a hard BA segment, and a flexible Cn segment [85]. The synthesis is described in Scheme 1. BA-C8 and BA-C10, which crystallize slowly, are ideal candidates for in-situ AFM crystallization studies. Their physical properties are summarized in Table 1.

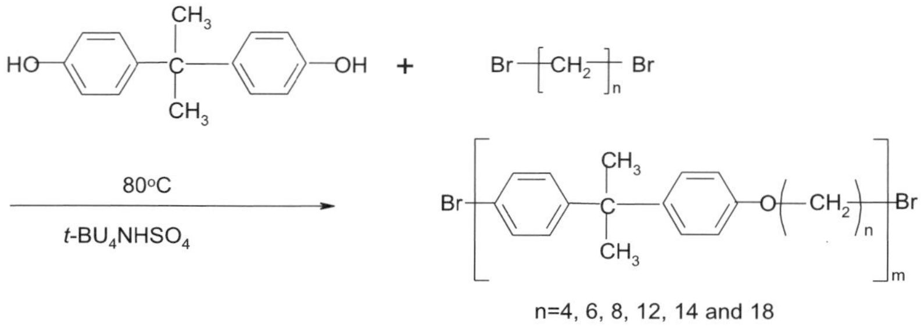

Scheme 1 Reaction and general structures of the BA-Cn polymers

Table 1 Results of the characterization of the BA-Cn sequential polymers [85]

Samples	T_g (°C)	T_m (°C)	$\overline{M}_w$ ×10^3	$\overline{M}_w/\overline{M}_n$
BA-C4	50.9	186.0	21.1	2.4
BA-C6	31.0	97.4	24.1	1.7
BA-C8	6.3	84.3	9.5	1.7
BA-C10	0.3	88.5	13.8	2.0
BA-C12	−5.9	88.3	16.2	2.1
BA-C14	−6.7	86.9	29.6	3.5
BA-C18	−4.1	81.8	28.6	3.5

2
Crystallization Processes

2.1
Homogeneous Nucleation—Birth of Primary Nuclei

Simple thermodynamics theory has predicted that at the on-set of crystallization, embryos have to form. As an embryo increases in size, its free energy increases until it reaches a critical size beyond which the volume-to-surface ratio becomes large enough such that the decrease in the enthalpy is larger than the increase in the surface energy. Below the critical size, the embryo can disintegrate and disappear. Beyond this size, the embryo becomes a stable nucleus and growth can continue, as shown in Fig. 2. Although the appearance and disappearance of the embryo have been predicted for a long time, these phenomena were never observed directly. In the last few years, such observations have been possible using AFM [59, 61, 73].

Li et al. [59] and Lei et al. [61] observed embryos of BA-C8 appearing as 10-nm dots and some disappearing several minutes later. Figure 3 shows two continuous AFM phase images of the same area. The time interval between

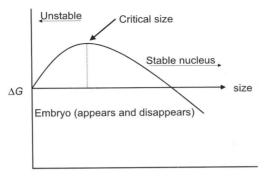

Fig. 2 Gibbs free energy of formation of a nucleus as a function of its size

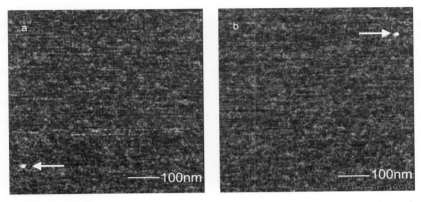

Fig. 3 Phase images obtained on a BA-C8 film at room temperature. **a** An embryo; **b** the embryo in **a** disappeared and a new embryo appeared at a different location [61]

the two images is 10.6 min and the image size is 600 nm. It can be seen clearly that an embryo appears as a 10-nm diameter dot in the lower left-hand corner of Fig. 3a. After 10.6 min, this embryo cannot be found (Fig. 3b); however, a new embryo appears in the upper right-hand corner of the image. These results demonstrate that embryos below a certain critical size can disintegrate, as predicted by thermodynamics. However, the accurate measurement of this critical size cannot be obtained with AFM.

However, one concern about the above studies is that the observed nucleation was a result of the influence from the AFM tip. Studies have been performed to investigate the nucleation induced by AFM tips during the crystallization experiments [48, 55, 77–79]. Pearce and Vancso investigated the possibility of tip-induced nucleation during crystallization of PEO near the melting point [48, 77]. They concluded that the tip did not induce nucleation. A similar conclusion was drawn by Godovsky and Magonov during their study of PE crystallization using AFM [55]. However, tip-induced nucleation was found to be possible in the study of crystallization of polycaprolactone (PCL) [78, 79]. The difference in the behavior between PEO and PCL was attributed to the difference in the shear force. High shear forces will cause chain alignment at the surface of the melt, as shown in Fig. 4. The lamellae are more or less aligned in the vertical direction. The tip scanning in the horizontal direction might cause the alignment of the polymer chains at the surface of the melt in the same direction. Hence, the growth direction of the lamellae is in the vertical direction, as shown in Fig. 4.

It is important to point out that in the studies of PCL [78, 79] and PEO [48, 77], the contact mode, which in general produces high shear forces during scanning, was used at temperatures near the melting points of the polymers. In experiments performed by Li et al. [59] and Lei et al. [61], the tapping mode was used and the experiments were performed at a high degree of undercooling ($\Delta T \sim 60\,^\circ\mathrm{K}$). During the experiments, the amorphous areas around the growing crystallites were scanned for a long time without detection of nucleation.

The appearance and disappearance of embryos of the ether-soluble (ES) fraction of stereoblock PP were observed by Schönherr et al. [73]. Features 1, 2, and 3, which are clearly visible in Fig. 5a, disappear in Fig. 5b. A stable nucleus, for example, Feature B in Fig. 5a, develops into a lamella. These studies have clearly demonstrated that the *primary* nuclei of polymers can be observed using AFM.

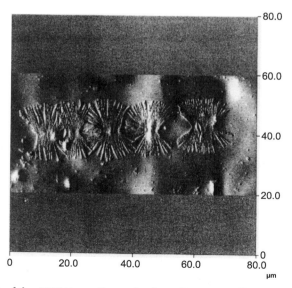

Fig. 4 The effect of the AFM tip on the nucleation of PEO crystallization in contact-mode AFM scanning. The micrograph exhibits a "defection contact-mode" image [79]

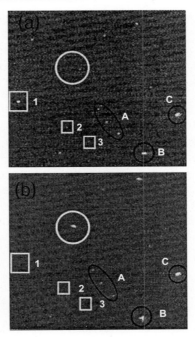

Fig. 5 Early stages of ES crystallization followed by hot stage AFM at 41 °C. The time lapse between the capture of the two phase images was 256 sec (image size = 3.3 mm × 3.0 mm) [73]

2.2
Development of the Founding Lamella

Once an embryo grows larger than the critical size as predicted by thermodynamics, the embryo can grow continuously at its two ends and develop into a single lamella, as shown in a series of images (c.f., Fig. 6). We can see an embryo as a round dot in the center of the image in Fig. 6a and the development of the embryo into a short lamella of approximately 60 nm in length in Fig. 6b. The appearance of this short lamella in Fig. 6b at the same location as the round dot in Fig. 6a provides solid evidence that the dot is an embryo. After about 60 min, as shown in Fig. 6c–f, this lamella grows in length to approximately 1 μm. This lamella, which originates from an embryo, is named as the founding lamella because all other lamellae that are present in the spherulite are its descendants.

The founding lamella grows into a lamellar sheaf through branching and splaying. The lamellae formed in the branching are referred to as the subsidiary lamellae. It is very interesting to find that the founding lamella did not branch until it grew to a fairly long length of about 1 μm (Fig. 6f,g). The results suggest that, during this period, the growth rate of the founding lamella

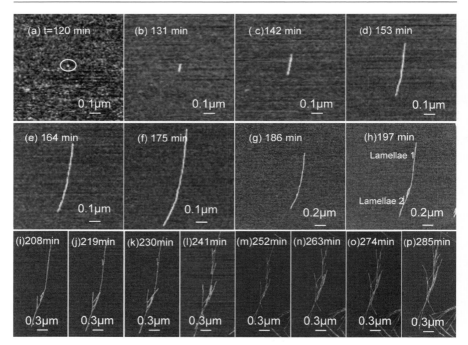

Fig. 6 A series of AFM phase images obtained on a BA-C8 film at room temperature. a An embryo; b a short lamella (founding lamella) developed from the embryo shown in a c–f. The growth of the founding lamella; g–p branching and splaying apart of the subsidiary lamellae [61]

is much faster than the rate of branching. Branching was found to occur on both folding surfaces of the founding lamella (c.f., Fig. 6i–p).

In order to determine whether the AFM tip induces the growth of the founding lamella in the direction normal to the fast scanning direction of the AFM tip, three AFM phase images were obtained at 25 °C, as shown in Fig. 7 [62]. The image area is 1.5 μm by 1.5 μm. The fast scanning direction is horizontal with respect to the image while the slow scanning direction is vertical. It is clear that the growth direction of these founding lamellae is independent of the fast scanning direction. These results indicate that with light tapping, the AFM tip scanning direction has no effect on the growth direction of the lamellae.

It is important to point out that lamellae with two orientations—edge-on and flat-on—are commonly observed by AFM, as shown schematically in Fig. 8. In general, AFM results reveal that edge-on and flat-on lamellae are dominant in thick and thin films, respectively. The founding and other edge-on lamellae of BA-C8 shown in Figs. 6 and 7 were formed on a film with a thickness of about 300 nm. Large flat-on lamellae surrounded by edge-on lamellae are seen on a BA-C10 film with a thickness of less 50 nm, as shown in

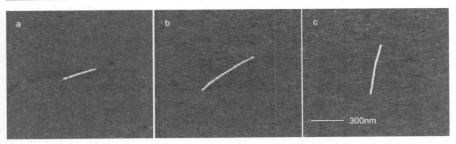

Fig. 7 AFM phase images showing founding lamellae of BA-C8 with different growth directions [62]

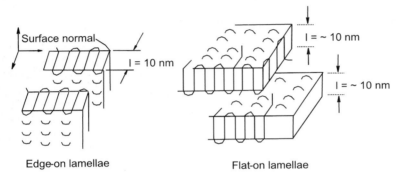

Fig. 8 Schematic showing edge-on and flat-on lamellae

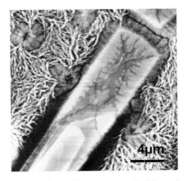

Fig. 9 Phase image showing flat-on lamellae surrounded by edge-on lamellae on a BA-C10 film with a thickness of less than 50 nm [64]

Fig. 9. The crystals in the center area are layer-packed, flat-on lamellae. They were crystallized at 75 °C for 90 h.

An explanation based on simple thermodynamics of the preferred orientation of edge-on and flat-on lamellae on thick and thin films, respectively, was given by Wang et al. [80]. As shown in Fig. 10, a primary rectangular nucleus of size $a \times a \times \ell$ can assume either edge-on or flat-on orientation on a sub-

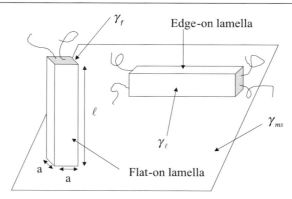

Fig. 10 Schematic showing the edge-on and flat-on nuclei on a substrate

strate. The free energy of formation for the edge-on primary nucleus, ΔG_{edge}, is given by

$$\Delta G_{\text{edge}} = -a^2 \ell \Delta G_b + 2a^2 \gamma_f + 3a\ell\gamma_\ell + a\ell(\gamma_{\text{cs}} - \gamma_{\text{ms}}) \tag{1}$$

where ΔG_b is the free energy change per unit of crystalline material formed, γ_f is the interfacial energy between the folding surface and the melt, γ_ℓ is the interfacial energy between the lateral plane and the melt, γ_{cs} is the interfacial energy between the crystal and the substrate and γ_{ms} is the interfacial energy between the melt and the substrate. By taking the derivative of ΔG_{edge} with respect to a and ℓ and setting them equal to zero, the critical dimensions (a_c and ℓ_c) for an edge-on nucleus attached to a substrate can be obtained by

$$a_c = \frac{3\gamma_\ell + \gamma_{\text{cs}} - \gamma_{\text{ms}}}{\Delta G_b}; \tag{2}$$

$$\ell_c = \frac{4\gamma_f}{\Delta G_b}. \tag{3}$$

The free energy of formation for a flat-on nucleus attached to a substrate, ΔG_{flat}, is

$$\Delta G_{\text{flat}} = -a^2 \ell \Delta G_b + 4a\ell\gamma_f + a^2(\gamma_f + \gamma_{\text{cs}} - \gamma_{\text{ms}}). \tag{4}$$

The critical dimensions for the flat-on nucleus are

$$a_c = \frac{4\gamma_\ell}{\Delta G_b}; \tag{5}$$

$$\ell_c = \frac{2(\gamma_f + \gamma_{\text{cs}} - \gamma_{\text{ms}})}{\Delta G_b}. \tag{6}$$

ΔG_b can be calculated using the following equation:

$$\Delta G_b = \frac{\Delta H_b \Delta T}{T_m^o} \tag{7}$$

where ΔH_b is the enthalpy of melting per unit volume of crystal, T_m^o is the equilibrium melting point of a polymer, and $\Delta T = T_m^o - T_c$ (T_c is the crystallization temperature). Assuming $\gamma_f = 0.093$ J/m², $\gamma_\ell = 0.01$ J/cm², $\Delta H_v = 2.090$ kJ/m³, $\Delta T = 60\,°$K, and $T_m^o = 400\,°$K, ΔG_{flat} and ΔG_{edge} were calculated and the results are shown in Table 2.

The results shown in Table 2 reveal that edge-on primary nuclei are preferred if $\gamma_{cm} - \gamma_{sm} < 0$. In thick films, the growth of the lamellae will most likely assume the same orientation as the primary nuclei formed at the interface. As a result, the authors of [80] argued that edge-on lamellae should be observed on thick films by AFM. However, the appearance and disappearance of embryos have been observed by AFM. These embryos have to be formed at the surface because if they were grown from the edge-on nuclei formed at the polymer-substrate interface, they would have been large lamellae and would not disappear at the surface. When a nucleus is formed at the surface of a thick film, the edge-on orientation is preferred because the presence of the lateral plane of the crystal at the surface minimizes the surface energy of the system due to the fact that γ_ℓ is much smaller than γ_f.

However, in thin films, the growth of the edge-on lamellae is limited by the film thickness (h). The formation of many edge-on lamellae on a thin film cre-

Table 2 Calculated values for ΔG_{flat} and ΔG_{edge} [80]

$\gamma_{cs} - \gamma_{ms}$, $\times 10^{-3}$ J/cm²	ΔG_{flat}, J	ΔG_{edge}, J
0	1.7×10^{-19}	1.8×10^{-19}
-10	1.5×10^{-19}	-6.5×10^{-19}
-20	1.3×10^{-19}	-7.1×10^{-19}

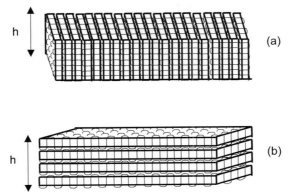

Fig. 11 Lamellar orientation in the ultrathin films: **a** many interfaces are created if the edge-on orientation is adapted; h is the film thickness; **b** few interfaces are created in the flat-on orientation

ates many interfaces between the edge-on lamellae, thus increasing the free energy of the system, as shown in Fig. 11. Consequently, continuous flat-on lamellar crystals are preferred on thin films. Although the thermodynamics model can explain the lamellar orientations in films with various thicknesses, the assumption that the interfacial energy between the crystal fold surface and the substrate is the same as the interfacial energy between the crystal lateral plane and the substrate is questionable.

3
Development of the Lamellar Sheaf

3.1
Branching of Lamellae

The formation of a lamellar sheaf, which is often referred to as a hedrite, as shown in Fig. 6p, is due to the branching and splaying apart of subsidiary lamellae. It is an intermediate and complex structure with a number of centrally connected lamellae. At the center, the lamellae are more-or-less parallel. For example, Beekmans et al. showed that hedrites of PCL with a high degree of complexity were developed from the melt upon lowering the crystallization temperature to approximately 56 °C [81]. A series of phase images of a hedrite in the edge-on orientation was acquired in the tapping mode at different time intervals (c.f., Fig. 12).

In the initial stage of the crystallization, the formation of a skeleton of dominant lamellae of equal widths separated by the melt is clearly visible as shown in Fig. 12a. The onset of branching is also visible in Fig. 12a. As the crystal grows, the hedrite becomes more asymmetrical with respect to the central dominant lamellae because it is tilted with respect to the surface (c.f., Fig. 12b–d). The dynamics of this space filling can clearly be observed in Fig. 12c,d. The subsidiary lamellae originating from the edge of the skeleton eventually develop a dominant character.

The mechanism of branching of lamellae has been the subject of intensive studies for many years. Bassett et al. investigated the lamellar morphologies of melt-crystallized i-PS and i-PP in detail [1, 5, 11, 27]. They proposed that the skeleton of a spherulite is established by individual dominant lamellae that branch and splay apart, leaving interstices to be filled by later-crystallizing subsidiary lamellae. The pressure build-up by molecular cilia between the dominant lamellae induces lamellar splaying. Lamellar branching is mainly induced through giant screw dislocations. In addition, they concluded that the subsidiary lamellae contain shorter molecules on average than do the dominant ones. At the same time, Norton and Keller studied the basic morphology of melt-crystallized i-PP [6, 7, 9]. Their investigations focused on five different spherulitic types, as identified by OM. Their results

Fig. 12 AFM phase images of PCL hedrite growing at 56 °C; the images in (**b**), (**c**) and (**d**) correspond to elapsed times of 244, 290, and 551 min, respectively, with respect to the image in (**a**) [81]

suggested that each spherulitic type is characterized by the arrangement of its constituent lamellae in terms of orientation, habit type, and crystal structure. Lamellar cross-hatch, namely lamellar branching, involves some form of "pure" epitaxy (i.e., a process of the deposition of one layer of helices with the same chirality as the previous layer; this deposition is preferred when isochiral helices are tilted about 80° to the helices in the (010) face of a parent lamella, resulting in the cross-hatched lamellar structures).

Figure 13 shows a flat-on lamellar crystal of BA-C8 formed at 75 °C on a film with a thickness of 200 to 250 nm. Spiral growth of crystals frequently occurs on the fold surface of the crystals. As shown in the schematic drawing (c.f., Fig. 13), when a screw dislocation emerges, polymer chains can attach on the ledge-like surface. After attachment of the chains, the next step is the chain folding into the crystal lattice. Then the growth proceeds in a spiral manner about the axis of the screw dislocation and finally develops to form this spiral structure. From the section analysis of the AFM height image, it

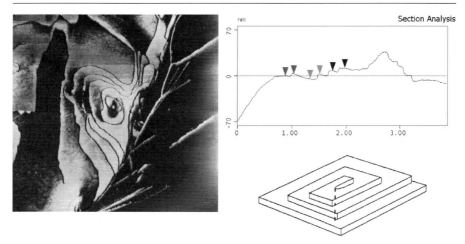

Fig. 13 Phase image showing flat-on lamellae of BA-C8 with a spiral structure

can be easily seen that the height profile of these terraces creates the different altitudes of the crystal structure.

The study of the growth processes of lamellae and spherulites of BA-Cn using AFM suggested that there are other possible mechanisms for lamellar branching [59–67]. Figure 14 shows the formation and the growth of two small nuclei (marked with A and B) near their parents. The newly formed nuclei grew and became subsidiary lamellae. A lamella breeds more lamellae as a result of the growth of these nuclei into subsidiary lamellae. Because some of the nuclei formed at a certain distance away from the parent lamellae, screw dislocations may not be the cause of branching unless the screw dislocations were formed at the subsurface level and grew to the surface. During crystallization, it is possible that a chain is partially trapped in a parent lamella. The reduced mobility of the trapped segments (a loop or cilium) induces the formation of a nucleus near the parent lamella, as shown schematically in Fig. 15. This phenomenon was called induced nucleation [61]. The nucleus formed in this process is the induced nucleus. It is interesting to see if an induced nucleus under a certain critical size will disappear like the primary nucleus. Figure 16a shows an induced nucleus near its parent lamella [82]. The induced nucleus slowly fades away as time increases. Finally, the nucleus disappears, as shown in Fig. 16b. On the basis of these results, it is concluded that an induced nucleus, which is formed near the parent lamella, could disintegrate when it is smaller than a certain critical size. These findings are similar to previous results showing that a primary nucleus smaller than a certain critical size is unstable and could disintegrate [59, 61, 73]. Because some of the induced nuclei were formed at a certain distance away from the parent lamellae, Lei et al. [61] and Wang et al. [82] proposed that screw dislocations might not be the cause of branching unless the screw disloca-

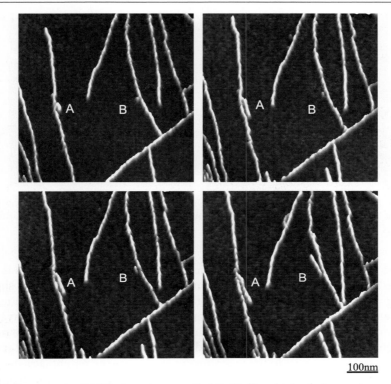

Fig. 14 A sequence of AFM phase images of a BA-C8 film obtained at room temperature showing the formation and growth of induced nuclei. The time interval between each consecutive image is approximately 5.8 min [61]

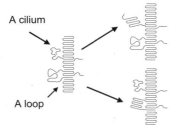

Fig. 15 Schematic showing the proposed mechanisms of induced nucleation

tions were formed at the subsurface level and grew to the surface. If a nucleus is formed at a subsurface level, this nucleus, which grows and appears at the surface, should be of significant size and would not disappear.

The appearance and disappearance of some of the induced nuclei suggest that they are formed at the surface as a result of a trapped polymer chain in the parent lamellae. In addition when a polymer crystallizes very quickly, stresses

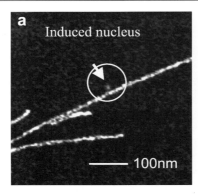 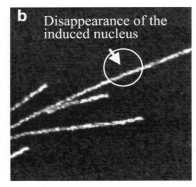

Fig. 16 AFM phase images of a BA-C8 film obtained at room temperature. **a** The presence of an induced nucleus near a parent lamella; and **b** the disappearance of the induced nucleus [82]

can be induced in its neighboring areas. The presence of high stresses is evidenced by the fact that the growing lamellae become highly curved, as shown in Fig. 17. At about 15 nm away from the highly curved lamella, three nuclei can be seen (marked with arrows). Because these nuclei are quite far away from the parent lamella, induced nucleation is unlikely to be the mechanism. Lei et al. [61] and Wang et al. [82] proposed that these nuclei were formed because of the stresses that exist in the neighboring areas near the parent lamellae. Although the branching mechanisms cannot be determined unequivocally based on the AFM, three possible branching mechanisms have been proposed: (1) Screw dislocations as suggested by Bassett; (2) induced nucleation as a result of a trapped polymer chain in the parent lamella; and (3) the stresses induced by the crystallization of a parent lamella in its nearby regions.

Fig. 17 AFM phase image obtained on a BA-C12 film. Three induced nuclei are observed at about 15 nm away from the parent lamella [83]

3.2
Branching at Different Temperatures

The proposed three mechanisms for branching are based on the defects generated during the crystallization process. It is likely that the rate of defect formation is related to the rate of lamellar propagation. Two parameters that are important in controlling the lamellar propagation rate are temperature and molecular weight. In this section the effect of temperature on lamellar branching is discussed.

Figure 18 shows the lamellar growth rate of a BA-C10 ($\overline{M}_W = 29.480$ g/mol) film with a thickness of 100 to 150 nm as a function of the crystallization temperature. The maximum lamellar growth rate occurs at around 55 °C. In order to relate the lamellar growth rate and branching of the lamellae, AFM images showing the lamellar growth at different temperatures were obtained. Figure 19 shows the AFM phase images of the lamellar branches developed at 35, 55, 70 and 75 °C. At about 35 °C, the branches are arranged fairly irregularly, as shown in Fig. 19a. The detailed statistical results show that the subsidiary lamellae have almost the same growth rates as the dominant lamellae. However, the lengths between the branch points and the tips of the subsidiary lamellae are approximately 200 to 500 nm shorter than the lengths between the branch points and the tips of the parent lamellae, as indicated by arrows in Fig. 19a. At 55 °C, it is difficult to distinguish between the dominant and subsidiary lamellae (c.f., Fig. 19b). In this case, the lengths between the branch points and the tips of the subsidiary lamellae are only 50 to 150 nm shorter than the lengths between the branch points and tips of the parent lamellae. At 70 °C, the number of lamellar branches is very small and the subsidiary lamellae are aligned orderly and almost parallel to the dominant

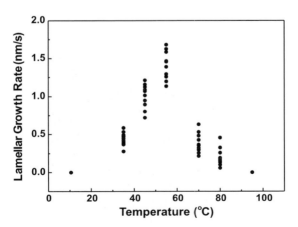

Fig. 18 Lamellar growth rate of BA-C10 as a function of crystallization temperature, determined by AFM [64]

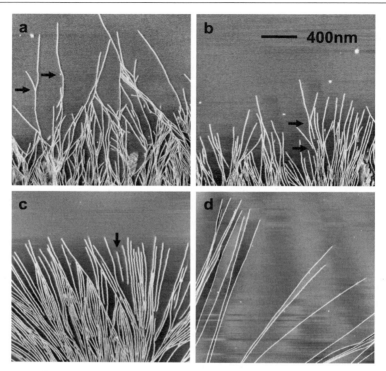

Fig. 19 AFM phase images of lamellae of BA-C10 crystallized at **a** 35 °C, **b** 55 °C, **c** 70 °C and **d** 75 °C [64]

lamellae as shown by the arrows in Fig. 19c. Some branching occurs almost at the tips of the dominant lamellae. It is difficult to determine whether they are generated from induced nuclei or from primary nuclei. Branching is not observed at 75 °C and the lamellae grow very straight, as shown in Fig. 19d. A comparison of the phase images of the lamellar branching in Fig. 19 indicates that the thickness of the lamellae grown at 70 and 75 °C appears to be larger than that of the lamellae developed at 35 and 55 °C.

Figure 20a is a schematic showing branching of a founding lamella where g is the average growth rate of the lamella and L_0 is the length between the lamellar tip and the location at which an induced nucleus appeared. The AFM results indicate that the growth rates of the founding lamella are quite uniform with time at a given crystallization temperature. Thus, the induction time, t_i, for the formation of an induced nucleus is defined as

$$t_i = \frac{L_o}{g}. \tag{8}$$

However, it is difficult to measure L_o, while in most cases, L_1 and L'_1, as defined in Fig. 20a, could be obtained easily when a subsidiary lamella grew to

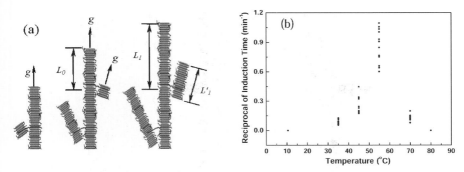

Fig. 20 **a** A schematic illustrating the branching of a founding lamella and **b** a plot of the reciprocal of induction time for subsidiary lamellae as a function of crystallization temperature [64]

a certain length. Thus, the induction time is given by

$$t_i = \frac{L_o}{g} = \frac{L_1 - L'_1}{g}. \tag{9}$$

The induction time was calculated using the measured values of L_1, L'_1, and g for the founding and subsidiary lamellae. The reported value is an average of five to ten measurements. The minimum in the plot of t_i as a function of the crystallization temperature was determined to be at about 55 °C, which is the same as the temperature at which the maximum lamellar growth rate occurs. As seen in Fig. 20b, induced nuclei hardly form when the crystallization temperature is lower than about 30 °C. Thus, the number of lamellar branches is quite small at this temperature. It is important to note that when the crystallization temperature is higher than 75 °C, induced nuclei also do not form. As a result, subsidiary lamellae are mostly absent and spherulites cannot form above this temperature. But other factors such as impurities can still promote the formation of *primary* nuclei. Hence, at this temperature, a heterogeneous nucleus can grow into a single crystal.

It is easy to understand that at low crystallization temperatures, the lamellar branching rate is low because the adjustment of the chain conformations is limited due to the immobility of these partially trapped chain segments. At high crystallization temperatures near the melting point, the driving force for secondary nucleation decreases, resulting in lower growth rates of the lamellae. The reduction in the induced nucleation rate at high temperatures is due to the fact that the increased thermal mobility of polymer chains and the small secondary nucleation rate provide enough time for the polymer chains to adjust their conformations to fold into a lamella with few protruding cilia or partially trapped chain segments. As a result, lamellar branching seldom occurs at high crystallization temperatures. A perfect edge-on lamellar single crystal without any branches develops at 80 °C ($T_m = 95$ °C) and

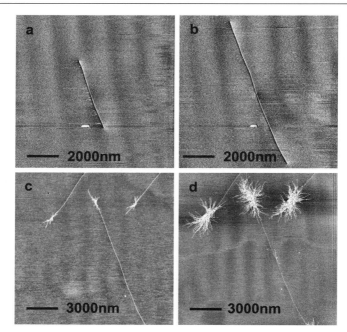

Fig. 21 In-situ phase images of a BA-C10 film showing the lamellar branching at different temperatures: **a** and **b** no branching at 80 °C; **c** and **d** branching largely at the tips of lamellae when the temperature is quenched to 35 °C [64]. The time interval between **a** and **b** was about 106 min

grows to 20 μm long, as shown in Fig. 21a,b. However, when the temperature is quenched to 35 °C, many lamellar branches appear at the growing tips of this single long lamella, as shown in Fig. 21c,d. The reason it was very difficult to form induced nuclei and lamellar branches at the center part of this single lamella is because this lamella developed at 80 °C with possibly no defects such as protruding cilia or trapped chain segments.

3.3
Propagation of Lamellae

AFM results on the propagation of lamellae show that the growth behavior of one lamella can be influenced by the presence of another lamella [60, 83]. The propagation of lamellae of BA-C8 has been studied and several interesting phenomena were observed. Figure 22 shows that the propagating lamellae can join each other. The joining sites of the lamellae are labeled J1, J2, and J3. The joining of two lamellae can occur at the tips of two growing lamellae (c.f., J1 in Fig. 22a,b). When the tips of two growing lamellae approach each other, the chain segments can fit into the two lamellae due to the imperfect crystal structure of the lamellar tips. The two lamellae resume growth and propagation after joining (c.f., J1 in Fig. 22c). The joining of two edge-on lamellae can also occur between the tips of a growing lamella and an existing lamella (c.f., J2 in Fig. 22a,b). This kind of joining can terminate the growth of one lamella without disturbing the growth rate of the existing lamella (c.f., J2 in Fig. 22b,c). Oppositely growing lamellae can also join together. This backward growth of a lamella may be terminated by another forward-growing lamella. When a forward-growing lamella meets a backward-growing lamella (c.f., J3 in Fig. 22a), they may join together (c.f., J3 in Fig. 22b). One of the very interesting observations is the joining of two lamellae initially propagating parallel to one another (c.f., P in Fig. 22b). The parallel growth of the two lamellae lasted only until they were about 1.0 µm in length. Then, the two lamellae grew closer to each other and joined together (c.f., P in Fig. 22c). Finally, the two lamellae separated and propagated again (c.f., P in Fig. 22d). It is reasonable to believe that polymer chains in the amorphous phase trapped between the parallel lamellae caused the joining of these two lamellae. The tension of the trapped polymer chains caused the joining of lamellae as the chain segments folding into the lattice of the lamellae pulls the two lamellae together, as illustrated in Fig. 23.

One of the most interesting phenomena observed is the bending of a lamella (c.f., B1 and B2 in Fig. 22). Lamella 1, as marked with B1 in Fig. 22,

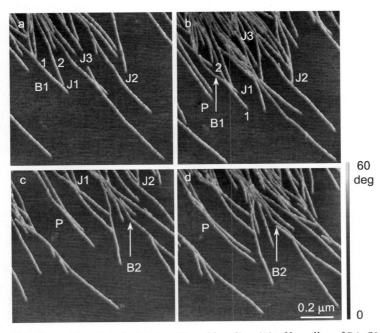

Fig. 22 Phase image showing the joining (J) and bending (B) of lamellae of BA-C8 during their growth and propagation [60]

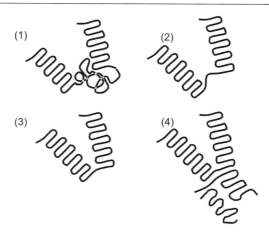

Fig. 23 Schematic showing the joining of two propagating lamellae [60]

becomes a bow-like lamella after joining with lamella 2, as shown in Fig. 22b. When lamella 2 approaches lamella 1, some loose loops of the polymer chain segments in lamella 1 can fit into the lattice of lamella 2 and the resulting tension can bend lamella 1. The bending of an edge-on lamella can also occur during its growth (c.f., B2 in Fig. 22c,d).

In another example, Hobbs et al. studied the lamellae growing in two neighboring shish-kebabs of PE, as shown in Fig. 24 [84]. One of the very interesting phenomena they observed is that two lamellae change their directions as they grow toward each other, resulting in a slight kink in one of the lamellae, as shown in Fig. 24b (labeled B). The results show that the growth direction of one lamella can be strongly influenced by the presence of another, although the mechanism by which this occurs is unclear. Their results also show that the contact of two lamellar tips can lead to the formation of a longer lamella that appears to be indistinguishable from a lamella that has grown as a single lamella because there is no sign of a join (c.f., area marked with A in Fig. 24c). On the basis of the AFM results on BA-C8 [60] and PE [84], several interesting phenomena can be found during lamellar propagation:

1. Two lamellae propagating in opposite directions can pass each other without changing their growth directions.
2. Two lamellae growing in opposite directions can join, forming one single lamella.
3. To avoid contact of two lamellae propagating in opposite directions, the growth directions of the lamellae can change, resulting in bending of the lamellae.
4. Two nonparallel lamellae propagating in a similar direction can join at a certain point. Then, they can propagate along different directions after joining. The join of these two lamellae can cause bending of the lamellae.

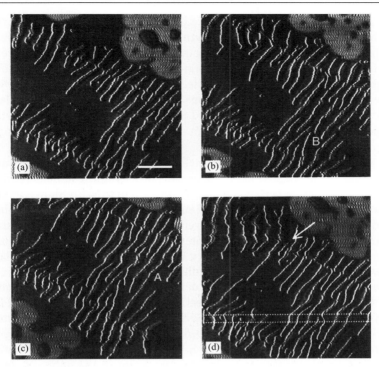

Fig. 24 A series of phase images showing the further growth of the PE shish-kebab structure. The gray scale represents a change in phase angle of 60°. The scan rate was 6.1 lines/sec. The scale bar refers to all the images and represents 300 nm. **a** Taken at 132 °C. **b** Taken at 131.5 °C; the B indicates a pair of lamellae that have changed direction to avoid joining. **c** Taken at 131 °C; the A indicates a pair of lamellae that have joined. **d** Taken at 130.5 °C; the *arrow* indicates a point on the extended chain backbone where a new nucleation event has occurred, and the *dotted lines* show the distorting effect of drift, in which all the lamellae on a series of scan lines are deformed [84]

5. Two parallel lamellae propagating in the same direction can grow toward each other and finally join together due to the tension of the polymer chains trapped between the two lamellae. After joining, these two lamellae can separate and propagate again.

These results reiterate the unique contributions of AFM in providing new insights into lamellar propagation.

3.4
Lamellar Growth Rate

The growth rate is important for any kinetic study. Many studies have been performed to measure the growth rate of spherulites using OM and recently

AFM. It has been predicted by theoretical models and observed experimentally that the overall spherulitic growth rate is constant. As for an individual lamellar growth rate, Hobbs pointed out that some lamellae initially grew forward faster than the overall growth rate of spherulites [59]. In addition, the growth rate was not constant. It varied for different lamellae, and even for the same lamellae at different locations [59]. AFM results show that a lamella might propagate faster in the forward direction than the overall growth rate and the propagation could slow down or even stop for a period of time and then restart again [60].

To study the growth rate of lamellae, their lengths were measured as a function of time at 22 °C [62]. The lengths of a founding lamella (lamella 1) and a subsidiary lamella (lamella 2) as a function of time are shown in Fig. 25. The growth rates of the founding and subsidiary lamellae were determined to be about 19 and 9 nm/min, respectively, from the slopes of the two straight lines in Fig. 25. It is important to note that the growth rate of the founding lamella is twice that of the subsidiary lamella because the founding lamella has two growing tips and the subsidiary lamella has only one. For convenience, the growth rate of lamellae is defined to be approximately 9.5 nm/min at 22 °C, using the growth in one direction only. The growth rate of all lamellae was found to be quite uniform. Such results seem to be contradictory to those of the earlier observations [59, 60]. It should be noted, however, that the earlier observations were made at different stages of the formation of spherulites (in the middle stage of the formation of the spherulites) while

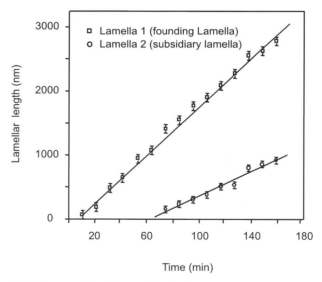

Fig. 25 Plots of the length of founding and subsidiary lamellae of BA-C8 as a function of time at 22 °C [62]

this observation was made at the initial stage of the formation of a lamellar sheaf. Together, these results clearly indicate that the growth rate of lamellae is different at different stages during the formation of the spherulites. The difference should reflect the influence of the concentration of developed lamellae on the growth rate of the lamellae.

At the middle stage of the formation of spherulites, there is a high concentration of growing lamellae. Because the concentration of the growing lamellae is not uniform across the film, the growth rate should also vary significantly even for the same lamella. However, at the initial stage of the formation of the spherulites, the concentration of trapped polymer chains is low and the time needed for the chain segments to adjust their conformation to fit into the lattice is affected mainly by the original orientation of the chains. Hence, the variation of the growth rate among the lamellae is small.

To explore the effect of temperature on the growth rate of the lamellae, the length of the lamellae was measured at two other temperatures (16 and 28 °C) and the results are shown in Fig. 26. The growth rate of the lamellae at 16 and 28 °C was measured to be 8.2 and 32.4 nm/min, respectively. The rate increases significantly as the temperature increases indicating that the growth rate of the lamellae at this temperature range is limited by the diffusion of the polymer chains.

The growth rate of lamellae at the later stage is not uniform. Figure 27 shows a sequence of AFM height and phase images of the lamellae at the growth front of a spherulite [60]. The time interval between each consecutive

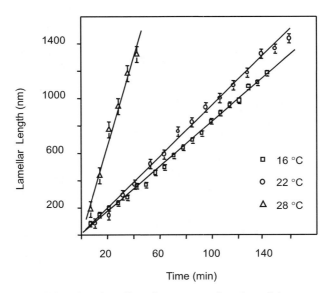

Fig. 26 The length of founding lamellae of BA-C8 as a function of time measured at three temperatures [62]

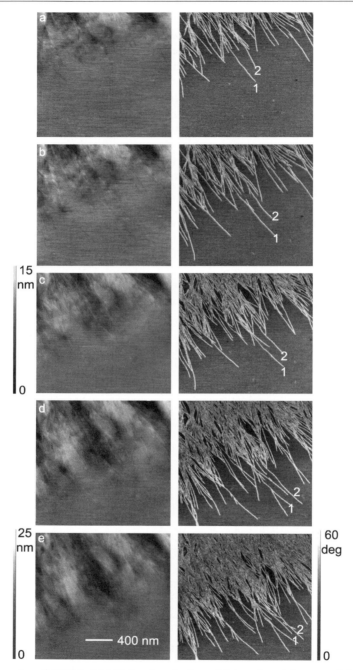

Fig. 27 A series of AFM height (*left column*) phase (*right column*) images of the growth front of a spherulite of BA-C8 in a consecutive time sequence [60]

image is approximately 9 min. The left column of Fig. 27 displays the AFM topographic images and the right column shows the corresponding phase images. The data scales of the AFM height images are 15 nm in Fig. 27a–c and 25 nm in Fig. 27d–e. As shown in the height images, the surface of the BA-C8 thin film becomes rough as random chain segments accommodate themselves into the lattice as the lamellae propagate. Individual lamella can be seen clearly in the phase image (right column of Fig. 27). The lamellae in the front of a spherulite do not grow at a constant rate. A lamella can propagate with various speeds at different times and locations (c.f., lamellae 1 and 2 in Fig. 27). At the beginning of this series of images, lamella 2 is behind lamella 1. But later as the growth rate of lamella 2 increases (c.f., Fig. 27c,d), lamella 2 surpasses lamella 1 (c.f., Fig. 27d). A similar phenomenon was observed in a film of poly(hydroxybutyrate-co-valerate) copolymer [58].

4
The Effect of Film Thickness on Lamellar Growth Rate and Morphology

It is known that film thickness can affect the orientation of the lamellae at the surface. As discussed in Sect. 2.2, in thin films, flat-on lamellae are preferred over edge-on lamellae because the growth of the edge-on lamellae is limited by the thickness of the film. The formation of many small edge-on lamellae increases the interfacial area of the system. As a result, flat-on lamellae are preferred. Film thickness, which has a significant effect on the mobility of the polymer chains in a film, can also affect the lamellar growth rate and morphology. Torres et al. applied a hard-sphere molecular dynamics methodology to calculate the effects of the polymer-substrate interaction and the thickness on the glass transition temperature (T_g) of polymer films [86]. Figure 28

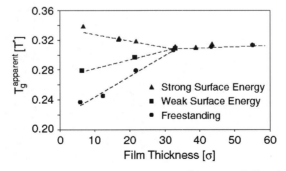

Fig. 28 Apparent glass transition temperature as a function of film thickness. The estimated uncertainty in the simulated T_gs is approximately four times the size of the symbols [86]

shows the apparent T_g as a function of film thickness (σ is the diameter of each monomer). For a free standing film, the glass transition temperature decreases significantly when the thickness decreases. The T_g of free-standing PS films with a thickness of 29 nm was measured to be 60 to 70 °K below its bulk T_g [87, 88]. A much smaller decrease in the T_g is expected for supported films with a weak polymer-substrate interaction. The T_g of supported PS films was reported be less than 10 °K below its bulk value [90]. An increase in the T_g of supported films is observed when the polymer-substrate interaction is strong. The polymer chains that are adjacent to the substrate with favorable interactions will have less mobility compared with those in the bulk. This argument is supported by experimental data showing that the glass transition temperature of the polymers increases as the film thickness decreases [72, 89–91]. Figure 29 shows the derived T_g for PEO films on oxidized silicon as a function of film thickness [90]. As the film thickness decreases, the T_g is found to increase as much as 20 °K above its bulk value.

A reduction in film thickness, which can increase or decrease the polymer chain mobility, has a significant impact on the crystalline morphology [72, 89, 92, 93] and also reduces the degree of crystallinity and lamellar growth rate. When a film is very thin (< 100 nm), dewetting of the polymer can occur, forming various interesting morphological patterns on the substrate. Schönherr and Frank studied the morphology and crystallization kinetics of isothermally crystallized PEO films with various thicknesses on oxidized silicon substrates [72]. In films with thicknesses larger than 1 µm, edge-on lamellae are predominantly present on the surface, as shown in Fig. 30a. When the thickness of the films is less than 300 nm, flat-on lamellae are the dominant structures

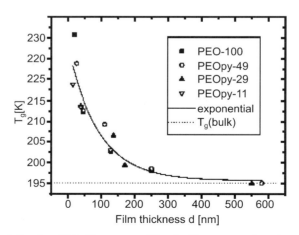

Fig. 29 Derived Tg for PEO films on oxidized silicon as a function of film thickness [90]. PEO-100 ($\overline{M}_W$ = 100 000 g/mole), PEOpy-49 ($\overline{M}_W$ = 49 340 g/mole), PEOpy-29 ($\overline{M}_W$ = 28 930 g/mole), and PEOpy-11 ($\overline{M}_W$ = 10 800 g/mole)

Fig. 30 TM-AFM phase images of PEOpy-49 films with various thicknesses on oxidized silicon: **a** 2.5 μm, **b** 110 nm, and **c** 15 nm [72]

on the surface and even the intentionally induced edge-on lamellae rotate and assume the flat-on orientation and continue to grow as flat-on lamellae. Figure 30b shows many flat-on lamellae with spiral growths developed from the screw dislocations. Films with thicknesses less than 15 nm tend to break up and form dendritic structures, as shown in Fig. 30c.

Kikkawa et al. studied the crystallization behavior of Poly[(s)-lactide] (P[(S)-LA]) from the melt at temperatures over 160 °C [93]. Their results show that the crystalline morphology is strongly influenced by the film thickness. Dendritic and hexagonal crystals were found, respectively, to form in thin films with thicknesses of 30 and 50 nm. AFM deflection images of a thin film with a thickness of about 30 nm, as shown in Fig. 31, were recorded in the contact mode isothermally at 165 °C after melting at 220 °C. A small crystal is detected at the central portion of the image after the temperature of the film reaches 165 °C for 5 min. Figure 31b–d show sequential AFM images of the P[(S)-LA] thin film at 165 °C. The initially formed narrow crystal, as shown in Fig. 31a, is likely to be an edge-on lamella that elongates and grows

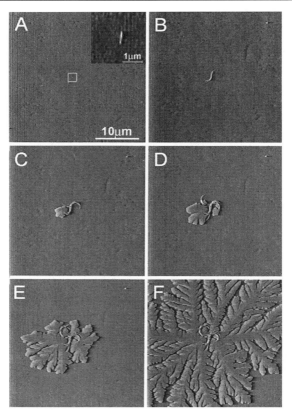

Fig. 31 AFM deflection images of P[(S)-LA] thin film with a thickness of about 30 nm, taken during isothermal crystallization at 165 °C on the AFM heating stage. In this sequence, the first frame (**a**) was taken at 7.5 min after the temperature reached 165 °C, and the following frames were taken at 15 (**b**), 22.5 (**c**), 30 (**d**), 50 (**e**), and 120 min (**f**), after reaching 165 °C [93]

to form an S-shaped structure (c.f., Fig. 31b). Flat-on lamellae are seen to form on one side of the narrow crystal and continue to grow (c.f., Fig. 31c). Another flat-on lamella is generated on the opposite side of the S-shaped narrow crystal, as shown in Fig. 31d. These flat-on lamellae grow with branching and finally form a dendritic morphology. It is clear from this example that flat-on lamellae are the preferred structure in thin films.

Figure 32 shows the continuous AFM deflection images of growing crystals in a (P[(S)-LA]) thin film with a thickness of about 70 nm obtained isothermally at 165 °C using tapping-mode AFM. After 5 min, some concave areas are detected on the amorphous surface of the thin film (c.f., Fig. 32a). In the concave area, edge-on lamellae emerge and grow with time. Many S-shaped and bow-shaped structures form on the surface after 38 min, as shown in

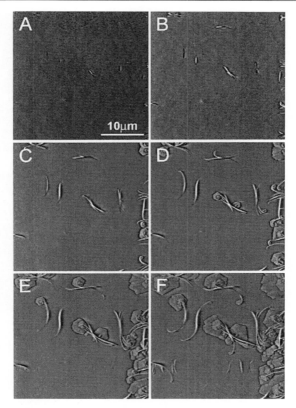

Fig. 32 Series of AFM deflection images showing flat-on and edge-on lamellae in P[(S)-LA] thin film with a thickness of 70 nm during isothermal crystallization at 165 °C. The first image (**a**) was started at 5 min after the temperature was stable at 165 °C. The following images were taken at 10 (**b**), 17 (**c**), 24 (**d**), 31 (**e**), and 38 min (**f**), respectively [93]

Fig. 32d. Flat-on lamellae are seen to develop at the tip of the edge-on lamellae. Instead of forming dendritic crystals as in the 30-nm film, the flat-on lamellae of this 70-nm film are hexagonal. The changes in the crystalline morphology from hexagonal to dendritic crystals when the film thickness decreases suggest that these morphological changes are caused by changes in the polymer chain mobility. In this thicker film, many more edge-on lamellae are present compared with the thinner 30-nm film, confirming that edge-on lamellae are preferred in thicker films.

The topographical AFM images of poly(ethylene vinyl acetate) films with various thicknesses ranging from 20 to 460 nm are shown in Fig. 33 [80]. The bulk-like spherulites are seen in the 460-nm film. In thick films, the surface morphology of the film is very similar to the bulk. As the thickness decreases to 152 nm, more small spherulites are observed. This is possibly due to the

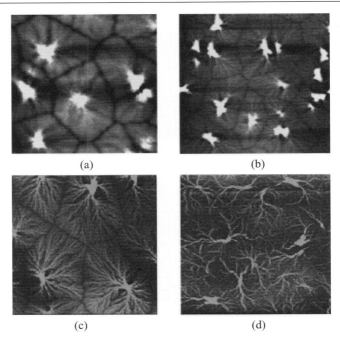

Fig. 33 AFM height images of the EVA films with thickness: **a** 460 nm; **b** 152 nm; **c** 70 nm, and **d** 20 nm. Scan size: 50 μm [80]

fact that a larger number of nuclei can be formed as a result of the slower crystallization growth rate. At the film thickness of 70 nm, the packing of the spherulites becomes more open. When the film thickness further reduces to 20 nm, hedrites appear and the spherulitic boundaries disappear, as shown in Fig. 33d.

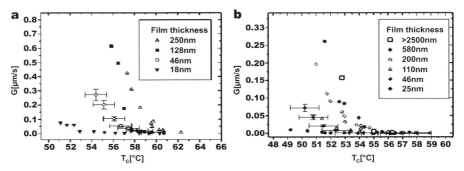

Fig. 34 a Growth rate G as a function of crystallization temperature T_c for different film thicknesses of PEO-100 (M_W = 100 000 g/mol). **b** G as a function of T_c for different film thicknesses of PEOpy-49 (M_W = 49 340 g/mol) [90]

The lamellar growth rate is undoubtedly controlled by the chain mobility of the polymers. Schönherr and Frank measured the lamellar growth rate of PEO and pyrene end-labeled PEO (PEOpy) with different molecular weights [90]. The lamellar growth rates were determined by measuring the width of the flat-on lamellae as a function of time. In films with a thickness greater than 200 nm, the thickness does not have a significant effect on the lamellar growth rate. However, in films with a thickness smaller than 200 nm, the lamellar growth rate decreases markedly with the thickness of the film, as shown in Fig. 34.

5
Formation of Spherulites

Figure 35 displays a series of AFM phase images obtained at 30 °C showing the development of a lamellar sheaf into a spherulite [62]. A lamellar sheaf finally develops into a spherulite through branching and splaying. Figure 35 also shows the formation of a pair of eyes at the center of the spherulite. A higher resolution phase image of the eyes for another sample obtained at 30 °C reveals that the eyes consist of only the flat-on lamellae, as shown in Fig. 36. One possible explanation is that as the edge-on lamellae start to propagate along the perimeter of the eyes, the strain energy increases due to the bending of the lamellae, as shown in Fig. 35d–f. In order to propagate in a lower energetic state, flat-on lamellae become the preferred orientation because they can fill up the space in the eyes without causing any bending of the lamellae. The diameter of the eyes typically ranges between 1 to 2 microns, as shown in Fig. 36.

Figure 37 is another series of phase images obtained at room temperature showing the formation of a spherulite from a lamellar sheaf and the contact of two spherulites. The primary embryo can form in the bulk and grow to the surface forming the angle-view lamellae. Through induced nucleation, the angle-view lamellae propagate to form edge-on lamellae at the surface and the edge-on lamellae spread across the film surface to form a spherulite. Figure 37 shows the growth process of the edge-on lamellae that are bred by the angle-view lamellae. It is noted that the size of the symmetric eyes at the center of the spherulite also grows with time. The growth rate of the eyes is slower than that of the edge-on lamellae, see Fig. 37a–f. The growth of the eyes is arrested upon contact with the edge-on lamellae around them (see Fig. 36).

These observations reveal that the periphery of the developing spherulites is not as smooth as that observed under OM. The growth front of the spherulite looks like a hedgehog and the lamellar stack develops into a spherulite shape only near the end of the growth, as shown in Fig. 37. Furthermore, at the early stage, the spherulite is not round in shape.

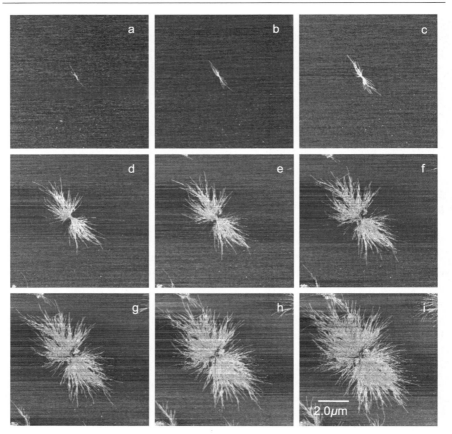

Fig. 35 Phase images showing the growth of a spherulite of BA-C8. The time interval between the images was 14.2 min except for the interval between (**c**) and (**d**) which was 53 min [62]

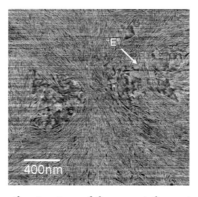

Fig. 36 Phase image shows the structure of the eyes at the center position of a spherulite of BA-C8 [62]

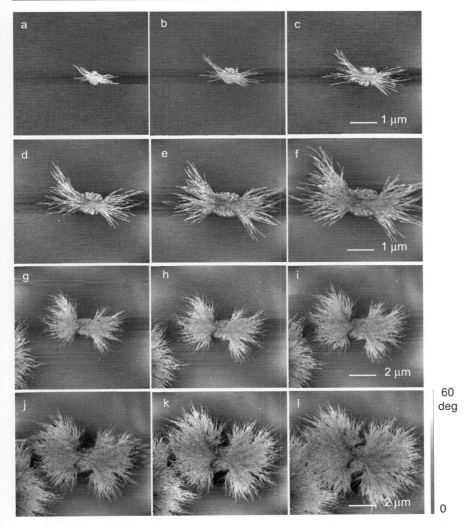

Fig. 37 A series of AFM phase images showing the formation of a BA-C8 spherulite [60]

6
Heterogeneous Nucleation

The previous sections described the growth process of spherulites induced by homogeneous nucleation. Heterogeneous nucleation, which can be identified by observation of lamellae growing in radial directions normal to the surface of the nucleus [17], was also studied. Figure 38 is a series of phase images obtained at room temperature showing heterogeneous nucleation and the spherulitic growth process [62]. The images, which are 10 µm × 10 µm

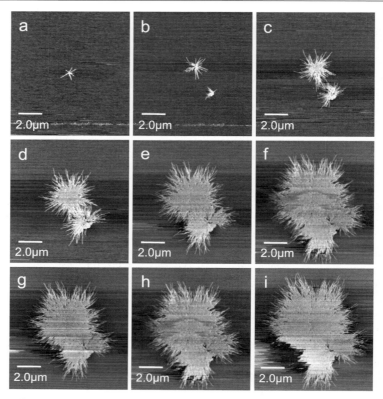

Fig. 38 Phase images showing the growth of a heterogeneously nucleated BA-C8 spherulite. The time interval between the images was 14.8 min [62]

in size, were taken at 30 °C. Figure 38a shows six edge-on lamellae growing normal to the surface of a nucleus. More edge-on lamellae continue to develop at the surface of the nucleus and grow outward radially (c.f., Fig. 38b). Subsidiary lamellae develop from induced nuclei (Fig. 38c). Branching and splaying continue and eventually a small spherulite is formed (Fig. 38d).

The structure of heterogeneously nucleated spherulites is different from the structure of homogeneously nucleated spherulites. First, in a heterogeneously nucleated spherulite, several lamellae can simultaneously develop on the surface of the nucleus. In a homogeneously nucleated spherulite, only one founding lamella is developed from a primary nucleus and all other lamellae are subsidiary lamellae originating from the founding lamella. In a heterogeneously nucleated spherulite, all lamellae grow into the melt in a direction normal to the surface of the nucleus. As a result, the spherulite has spherical symmetry starting even at the very early stage. No lamellar sheaf is seen. Consequently, the spherulites do not have a pair of eyes at their center.

Another spherulite formed as a result of homogeneous nucleation can be seen at the lower right side of the heterogeneously nucleated spherulite. These two spherulites grow and eventually impinge on each other as shown in Fig. 38g–i. It is interesting to note that the boundary is formed by the interpenetration of lamellae from the two spherulites. These results suggest that the spherulitic growth process can significantly affect the mechanical properties of semi-crystalline polymers.

7
Summary

During the last five years, AFM has demonstrated its capabilities as a remarkable tool for studying the crystallization processes of semi-crystalline polymers. It can provide real time and real space information in resolutions ranging from nanometers to micrometers. AFM results have shown that the appearance and disappearance of embryos as predicted by thermodynamics have been observed. The development of the founding lamella from a stable nucleus and the branching and splaying of the subsidiary lamellae from the founding lamellae, forming a lamellar sheaf, have been revealed by AFM in detail. In addition, the detailed development of spherulites from lamellar sheafs has been observed. With the development of hot-stage AFM, the kinetics of polymer crystallization can now be studied at different temperatures in real time. It is anticipated that many significant advances in the study of polymer crystallization will be made in the years to come.

Acknowledgements This work was supported by the Hong Kong Research Grants Council under grant numbers HKUST6176/02 and 600503, the National Science Foundation of China and the Hong Kong Research Grants Council Joint Research Scheme under Grant No. N_HKUST 618/01 as well as the Outstanding Youth Fund of the National Science Foundation of China.

References

1. Bassett DC (1981) Principles of Polymer Morphology. Cambridge University Press, Cambridge
2. Woodward AE (1988) Atlas of polymer morphology. Hanser Publisher, New York
3. Schultz JM (2001) Polymer crystallization: the development of crystalline order in thermoplastic polymers. Oxford University Press, Oxford
4. Strobl GR (1996) The physics of polymers: concepts for understanding their structures and behavior. Springer, Berlin Heidelberg New York
5. Bassett DC, Olley RH (1984) Polymer 25:935
6. Norton DR, Keller A (1985) Polymer 26:704
7. Keller A (1955) J Polym Sci 17:291

8. Keith HD, Padden Jr FJ (1959) J Polym Sci 39:101
9. Keller A, Sawada S (1964) Makromol Chem 74:190
10. Bassett DC, Hodge AM, Olley RH (1981) Proc R Soc A377:25
11. Bassett DC, Vaughan AS (1985) Polymer 26:717
12. Olley RH, Bassett DC (1989) Polymer 30:399
13. Bassett DC (1994) Phil Trans R Soc A 348:29
14. Li JX, Ness JN, Cheung WL (1996) J Appl Polym Sci 59:1733
15. Li JX, Cheung WL (1999) J Appl Polym Sci 72:1529
16. Phillips PJ, Andrews EH (1972) Polym Letters 10:321
17. Phillips PJ, Edwards BC (1975) J Polym Sci: Polym Phys 13:1819
18. Edwards BC, Phillips PJ (1975) J Polym Sci: Polym Phys 13:2117
19. Phillips PJ, Edwards BC (1976) Polym Lett 14:449
20. Phillips PJ, Sorenson D (1979) J Polym Sci: Polym Phys 17:521
21. Edwards BC, Phillips PJ, Sorenson D (1980) J Polym Sci: Polym Phys 18:1737
22. Phillips PJ, Sorenson D (1981) J Polym Sci: Polym Lett 19:585
23. Rensch GJ, Phillips PJ, Vatansever N (1986) J Polym Sci, Polym Phys 24:1943
24. Daxaben P, Bassett DC (1994) Phil Trans R Soc Lond A445:577
25. Bassett DC, Olley RH, Al Raheil IAM (1988) Polymer 29:1539
26. Bassett DC, Hodge AM (1981) Proc R Soc Lond A 377:25
27. Padden FJ, Keith HD (1959) J Appl Phys 30:1479
28. Bassett DC, Olley RH(1984) Polymer 25:935
29. Keith, HD, Padden FJ (1963) J Appl Phys 34:2409
30. Binning G, Quate CF, Gerber Ch (1986) Phys Rev Lett 66:930
31. Rugar D, Hansma P (1990) Physics Today 43:23
32. Miles JM (1997) Science 277:1845
33. Magonov SN (1993) Applied Spectroscopy Reviews 28:1
34. Sheiko SS (2000) Adv Polym Sci 151:61
35. Magonov SN, Reneker DH (1997) Ann Rev Mat Sci 27:175
36. Munz M, Capella B, Sturm H, Geuss M (2003) Adv Polym Sci 164:87
37. Magonov SN, Elings V, Whangbo MH (1997) Surf Sci 375:L385
38. Bar G, Thomann Y, Brandsch R, Cantow HJ, Whangbo MH (1997) Langmuir 13:3807
39. Bar G, Delineau L, Brandsch R, Bruch M, Whangbo MH (1997) Appl Phys Lett 75:4198
40. Bar G, Brandsch R, Whangbo MH (1999) Surf Sci 436:L715
41. Magonov SN, Godovsky YK (1999) American Laboratory 31:52
42. Höper R, Gesang T, Possart W, Hennemann OD, Boseck S (1995) Ultramicroscopy 60:17
43. Magonov SN, Cleveland J, Elings V, Denley D, Whangbo MH (1997) Surf Sci 389:201
44. Leclère Ph, Lazzaroni R, Brédas JL, Yu JM, Dubois Ph, Jérôme R (1996) Langmuir 12:4317
45. Patil R, Kim SJ, Smith E, Reneker DH, Weisenhorn AL (1990) Polym Comm 31:455
46. Snetivy D, Vancso GJ (1992) Polymer 33:432
47. Pearce R, Vancso GJ (1998) Polymer 39:1237
48. Pearce R, Vancso GJ (1998) J Polym Sci: Polym Phys 36:2643
49. Haeringen DTV, Varga J, Ehrenstein GW, Vancso GJ (2000) J Polym Sci: Polym Phys 38:672
50. Harbon HR, Pritchard RG, Cope BC, Goddard DT (1996) J Polym Sci: Polym Phys 34:173
51. Crämer K, Wawkuschewski A, Domb A, Cantow HJ, Magonov SN (1995) Polym Bull 35:457
52. Motomatsu M, Nie HY, Mizutani W, Tokumoto H (1996) Polymer 37:183

53. Ivanov DA, Jonas AM (1998) Macromolecules 31:4546
54. Trifonova D, Varga J, Vancso GJ (1998) Polym Bull 41:341
55. Godovsky YK, Magonov SN (2000) Langmuir 16:3549
56. Bartczak Z, Argon AS, Cohen RE, Kowalewski T (1999) Polymer 40:2367
57. Schultz JM, Miles MJ (1998) J Polym Sci: Polym Phys 36:2311
58. Hobbs JK, McMaster TJ, Miles MJ, Barham PJ (1998) Polymer 39:2437
59. Li L, Chan CM, Li JX, Ng KM, Yeung KL, Weng LT (1999) Macromolecules 32:8240
60. Li L, Chan CM, Yeung KL, Li JX, Ng KM, Lei YG (2001) Macromolecules 34:316
61. Lei YG, Chan CM, Li JX, Ng KM, Wang Y, Jiang Y, Li L (2002) Macromolecules 35:6751
62. Lei YG, Chan CM, Wang Y, Ng KM, Jiang Y, Li L (2003) Polymer 44:4673
63. Luo YH, Jiang Y, Jin XG, Li L, Lei YG, Chan CM (2002) Chinese Science Bulletin 47:1761
64. Jiang Y, Yan DD, Gao X, Han CC, Jin XG, Li L, Wang Y, Chan CM (2003) Macromolecules 36:3652
65. Luo YH, Jiang Y, Jin XG, Li L, Chan CM (2003) Macromolecular Symposia 192:271
66. Jiang Y, Zhou JJ, Li L, Xu J, Guo BH, Zhang ZM, Wu Q, Chen GQ, Weng LT, Cheung ZL, Chan CM (2003) Langmuir 19:7417
67. Jiang Y, Jin XG, Han CC, Li L, Wang Y, Chan CM (2003) Langmuir 19:8010
68. Schönherr H, Snetivy D, Vancso GJ (1993) Polym Bull 30:567
69. Nakamura J, Kawaguchi A (2004) Macromolecules 37:3725
70. Dubreuil N, Hocquet S, Dosière M, Ivanov DA (2004) Macromolecules 37:1
71. Schönherr H, Wiyatno W, Pople J, Frank CW, Fuller GG, Gast AP, Waymouth RM (2002) Macromolecules 35:2654
72. Schönherr H, Frank CW (2003) Macromolecules 36:1188
73. Schönherr H, Waymouth RM, Frank CW (2003) Macromolecules 36:2412
74. Koike Y, Cakmak M (2004) Macromolecules 37:2171
75. Imase T, Ohira A, Okoshi K, Sano N, Kawauchi S, Watanabe J, Kunitake M (2004) Macromolecules 36:1865
76. Hosier IL, Alamo RG, Lin JS (2004) Polymer 45:3441
77. Pearce R, Vancso GJ (1997) Macromolecules 30:5843
78. Beekmans LGM (2002) Morphology development in semi-crystalline polymers by in-situ scanning force microscopy. PhD Thesis, University of Twente
79. Vancso GJ, Beekmans LGM, Pearce R, Trifonova D, Varga J (1999) J Macromol Sci: Phys B38:491
80. Wang Y, Ge S, Rafailovich M, Sokolov J, Zou Y, Ade H, Luning J, Lustiger A, Maron G (2004) Macromolecules 37:3319
81. Beekmans LGM, Vancso GJ (2000) Polymer 41:8975
82. Wang Y, Chan CM, Ng KM, Jiang Y, Li L (2004) Langmuir 20:8220
83. Wang Y, Chan CM (unpublished work)
84. Hobbs JK, Humphris ADL, Miles MJ (2001) Macromolecules 34:5508
85. Li L, Chan CM, Ng KM, Lei YG, Weng LT (2001) Polymer 42:6841
86. Torres JA, Nealey PF, de Pablo JJ (2000) Phys Rev Lett 85:3221
87. Forrest JA, Dalnoki-Veress K, Stevens JR, Dutcher JR (1996) Phys Rev Lett 77:2002
88. Forrest JA, Dalnoki-Veress K, Dutcher JR (1997) Phys Rev E 56:5705
89. Van Zanten JH, Wallace WE, Wu WL (1996) Phys Rev E 53:R2053
90. Schönherr H, Frank CW (2003) Macromolecules 36:1199
91. Fryer DS, Nealey PF, de Pablo JJ (2000) Macromolecules 33:6439
92. Massa MV, Dalnoki-Veress K, Forrest JA (2003) Eur Phys J E 11:191
93. Kikkawa Y, Abe H, Iwata T, Inoue Y, Doi Y (2001) Biomacromolecules 2:940

Toughness of Neat, Rubber Modified and Filled β-Nucleated Polypropylene: From Fundamentals to Applications

Christelle Grein

Borealis GmbH, St.-Peter Straße 25, P.O. Box 8, 4021 Linz, Austria
christelle.grein@borealisgroup.com

1	Introduction	46
2	Some Preliminary Information	47
2.1	Main Differences between the α- and β-Modifications of PP	47
2.2	Quantification of the Amount of β-Modification	48
2.3	Assessment of Mechanical Performance	50
3	Toughness in β-Polypropylene	51
3.1	Influence of Intrinsic Characteristics	52
3.1.1	Influence of Molecular Weight	52
3.1.2	Influence of Polydispersity	54
3.1.3	Influence of a Matrix with Low Ethylene Content	55
3.2	Influence of Extrinsic Parameters	57
3.2.1	Influence of the Content of Nucleating Agent	57
3.2.2	Influence of the Type of Nucleating Agent	59
3.2.3	Influence of Processing Conditions	62
3.2.4	Influence of Stress-State	64
3.2.5	Influence of Test Speed and Temperature	66
4	Toughness of Filled β-Polypropylene	70
5	Toughness in Rubber-Toughened β-Polypropylene	73
5.1	Rubber-Toughened PP: A Brief Overview	73
5.2	Influence of Rubber Content	73
5.3	Influence of Rubber Molecular Weight	75
5.4	Deformation Map of a Specific β-Nucleated PP/EPR	76
5.5	Which Limiting Factor: Matrix or Rubber?	78
6	Specific Damage Mechanics in β-Nucleated PP	81
6.1	Some General Features	81
6.2	Some Experimental Findings	83
6.2.1	Enhanced Creation of Microvoids	83
6.2.2	Dynamic Mechanical Analysis (DMA) Evidences	85
6.2.3	Macroscopic Tensile Indicators	86
6.3	What Explains the Toughness of β-Modified Polypropylene?	89
6.3.1	Lamella Architecture	89
6.3.2	Additional Factors	90
6.3.3	The Controversial β-α-Phase Transformation	92

7	Some Applications of β-Nucleated Polypropylene	93
7.1	Fibers	93
7.2	Glass-Fiber Reinforced Polypropylene	94
7.3	Thermoforming	94
7.4	Microporous Films	95
7.5	Piping Systems	96
8	Conclusion	98
	References	99

Abstract This review highlights several aspects of the toughness of β-nucleated polypropylene (PP). The focus is on dynamic fracture properties, a topic which is largely documented in the literature. The role of intrinsic parameters like molecular weight, polydispersity and matrix randomization has been discussed, that of extrinsic factors like stress-state, processing conditions, test speed and temperature illustrated. Under defined conditions, the toughness is also defined by the content and spatial distribution of the β-nucleating agent. The increase in fracture resistance is more pronounced in PP homopolymers than in random or rubber-modified copolymers. In the case of sequential copolymers, the molecular architecture inhibits a maximization of the amount of β-phase; in that of heterophasic systems, the rubber phase mainly controls the fracture behavior. The performance of β-nucleated PP has been explained in terms of smaller spherulitic size, lower packing density and favorable lamellar arrangement of the β-modification which induce a higher mobility of both crystalline and amorphous phases. The damage process is accompanied by numerous microvoids, the development of which has been utilized for breathable films. Other interesting application segments are fibers, glass-fiber reinforced PP, thermoformable grades and piping systems.

Keywords Deformation mechanisms · Nucleation · Polypropylene · Toughness · β-Crystalline structure (or β-modification · β-Phase)

Abbreviations

α-PP	α-modification of isotactic polypropylene
β-PP	β-modification of isotactic polypropylene
γ-PP	γ-modification of isotactic polypropylene
α_c	high temperature relaxation of PP (DMA)
β	angle relaxation at 0 °C of PP (DMA)
β_1	first β-phase during DSC heat scan
β_2	second β-phase during DSC heat scan
γ	high temperature relaxation of polypropylene (DMA)
ε	strain
$\varepsilon_{cav}/\varepsilon$	cavitational contribution to strain
ε_{break}	elongation at break
v_i	activation volume of the element motion unit for the process i
σ_y	yield stress
θ	angle
ΔH	melt enthalpy (DSC)
$\Delta H(\alpha)$	melt enthalpy (DSC) of α-PP
$\Delta H(\beta)$	melt enthalpy (DSC) of β-PP

$\Delta H_{100\%}(\alpha)$	melt enthalpy (DSC) of 100% crystalline α-PP
$\Delta H_{100\%}(\beta)$	melt enthalpy (DSC) of 100% crystalline β-PP
ΔH_i	activation energy of the plastic flow for the mono-activated process i
Δ_{py}	difference between the plateau stress and the yield stress
a	crack length
	length of the a-parameter of a unit cell
b	length of the b-parameter of a unit cell
c	length of the c-parameter of a unit cell
d	displacement
$d\varepsilon/dt$	strain rate
dF/dt	initial slope of the force-time curve
dK/dt	crack tip loading rate
$f(a/W)$	geometrical dimensionless factor
p_C	critical pressure
t	time
v	test speed
w_e	specific essential work of fracture
B	thickness
Cw_p	specific non-essential work of fracture
D_w	average particle size in weight
E	Young modulus, stiffness
E_{limit}	elasticity limit
F	force
F_{max}	maximum of force
G_α	growth rate of α-phase
G_β	growth rate of β-phase
G_C	critical energy release rate (LEFM)
G_D	dynamic fracture resistance (LEFM)
G_{ini}	initiation energy
G_{plast}	plastic energy
G_{tot}	fracture energy
H	height of a crystalline peak (WAXS)
IS	impact strength
IV	intrinsic viscosity
J_{Id}	resistance to fracture (Integral J)
K_β	β-content
K_C	(critical) stress intensity factor (LEFM)
K_D	dynamic fracture toughness (LEFM)
L	lateral size of lamellae
LP	lamellar long period
M_n	number average molecular weight
M_w	weight average molecular weight
MFR	melt flow rate
MWD	molecular weight distribution
NIS	notched impact strength
dNIS	double-edged notched impact strength
PI	polydispersity
S	order parameter of the β-phase
T	temperature
T_c	crystallization temperature

T_{db}	temperature at which ductile-brittle transition occurs
T_g	glass transition temperature
T_m	melting temperature
V_c	cooling rate
W	width
ZC	size of diffuse whitened zone (on broken CT specimen)
ZI	size of intense whitened zone (on broken CT specimen)
BSE	back-scattering electron mode
CT	compact tension
DMA	dynamic mechanical analysis
DSC	differential scanning calorimetry
EPR	ethylene-propylene rubber
EWF	essential work of fracture
LEFM	linear elastic fracture mechanics
PP	polypropylene
PP/EPR	blend of PP and EPR, ethylene-propylene block copolymer
PTT	phase transformation toughening
RCP	rapid crack propagation
RuO_4	ruthenium tetraoxyde
SCG	slow crack growth
SEM	scanning electron microscopy
TEM	transmission electron microscopy
WAXS	wide angle X-ray scattering

1
Introduction

Among the three known crystalline structures (α, β, γ) of isotactic polypropylene (PP), the β-modification is certainly the most fascinating one. While the stable α-structure develops under standard process conditions, the occurrence of the β-form has to be forced (i) by directional crystallization in a temperature gradient field [1–3]; (ii) by shear-induced crystallization [4–12]; or (iii) by the addition of specific nucleating agents [13–31]. This latter technique is preferred at the industrial scale.

The most commonly used nucleation agents are γ-quinacridone (Permanent Red E3B) [13, 14, 21, 32–43], calcium pimelate or suberate (whichever production method is used) [19, 23, 26, 28, 42, 44–47, 49–58] and N,N-dicyclohexyl-2,6-naphthalene dicarboxamide produced by New Japan Chemicals—better known as NJ Star NU-100—[17, 24, 27, 59–69]. The reasons for their success are (i) their ease of production; (ii) their stability towards humidity and heat; (iii) their favorable organoleptics; (iv) their price; and (v) provided that their concentration and processing have been optimized, their high nucleating ability with the formation of more than 90% of the β-modification (as measured by X-ray diffraction). This efficiency is thought to be related to their crystallographic characteristics. Nucleating agents exhibiting a structural periodicity with a spacing close to $c/2 = 0.65$ nm, the axis repeat distance of the PP-helix, are suspected to enable epitaxial growth of the β-PP lamellae [27, 28, 43]. A detailed list of β-nucleating agents can be found in the thorough paper of Varga [70].

It has been widely reported that β-nucleated PP exhibits a superior impact performance as compared to their non-nucleated or α-nucleated homologues [32, 36, 41, 50, 53, 63, 64, 71–84]. A review of the state of the art, including our own partially unpublished investigations, will show the dominating influence of the PP characteristics on the fracture resistance of β-nucleated grades. The decisive effect of the strain-rate, temperature, type of loading and processing on the resulting toughness will also be discussed. Moreover, the specific damage mechanisms associated with the β-modification will be commented upon based on the latest developments. Besides their improved impact resistance, we will show how some of the outstanding characteristics of β-nucleated PP have been successfully converted into commercial grades. Special focus will be on pipe, microvoided films and thermoforming applications.

2
Some Preliminary Information

2.1
Main Differences between the α- and β-Modifications of PP

In both α- and β-modifications, the chains are packed in the lattice in the form of right- or left handed $2 \times 3_1$ helices, but their crystalline structure differ. The α-modification is monoclinic with $a = 6.61$–6.66 Å, $b = 20.73$–20.98 Å, $c = 6.495$–6.53 Å and $\beta = 98.5$–$99.62°$ [85–89]. Depending on the packing of its "up" and "down" helices (which accounts for the orientation of the methyl groups with respect to the chain axis), its space group is either C2/c (random packing of – CH_3 group) or P21/c (alternating packing of – CH_3 group) [88–91]. This latter, typical for low undercooling or annealing at elevated temperatures, is the most stable one [90–92]. The α-form is characterized by auto-epitaxial growth of lenticular lamellae, elongated in the [100] direction, giving rise to the so-called "cross-hatched" morphology in which two distinct populations of lamellae are arranged with their long axes roughly perpendicular locally [93, 94]. The induced α-spherulitic structures differ from those of the β-form in their size (50–100 μm for the non-nucleated α-form towards 1–10 μm for the β-phase) and in their birefringence, making it possible to distinguish both modifications in an optical microscope with polarized light. The β-spherulites often exhibit banded textures, as a result of the relatively broad β-lamellae (about 30 nm) which form coplanar stacks whose plane tends to twist about the growth direction [93,

95]. From a nanoscale point of view, β-PP crystallizes in a hexagonal form with $a = b = 11.01$ Å and $c = 6.49$ Å (the space group is probably P3$_1$) [86, 96–98]. It belongs to a class of polymers which displays a "frustrated" structure, referring to the fact that the helices in the unit cells have different azimuthal settings [37, 99–101]. The relative amounts of α- and β-forms in a sample are controlled by intrinsic molecular parameters and extrinsic factors. They will be discussed in detail in Sect. 3.2.

2.2
Quantification of the Amount of β-Modification

The unique proper way to quantify the amount of β-modification is based on the X-ray diffraction pattern of a PP sample (Fig. 1). Two peaks stand for the β-crystalline phase in a wide angle X-ray scattering (WAXS) trace: that at $2\theta = 14.2°$ representing the (300) plane and that at $2\theta = 21°$ accounting the (301) plane. The content of the β-form is determined by the widely accepted empirical equation of Turner-Jones [86]:

$$K_\beta = \frac{H(300)}{H(300) + H_\alpha} \quad (1)$$

where $H(300)$ is the height of the β 300 peak at $2\theta = 16.2°$ and H_α is the sum of the heights of the α peaks at $2\theta = 14.2°$, $17°$ and $18.8°$ which correspond to the respective (110), (040) and (130) planes. It is also possible to derive an order parameter S for the β phase from [102]

$$S = \frac{H(300)}{H(300) + H(301)} \quad (2)$$

where $H(301)$ is the height of β 301 peak at $2\theta = 21°$. The higher S, the higher the order of the β-phase.

In practice, differential scanning calorimetry (DSC) is often used in laboratory routine as a rapid and quite reliable alternative to WAXS to estimate the content of the β-crystalline structure in a specimen. It can, however, not be considered as an exact method since melting and re-crystallization of the β-form during the scan is suspected, leading to a β/α phase transformation, which might at least be partly responsible for the endothermic peak at around 160–170 °C, characteristic for the α-modification [70]. In samples submitted to a classical scanning program (+ 10/ − 10/ + 10 K min^{-1}), two endothermic peaks attributed to the β-form can occur (Fig. 2). Their melting temperature, T_m, taken as the maximum of the peaks, depends mainly on the cooling rate of the sample. While the first peak is always present and considered to reflect the melting range of the more or less disordered β-PP (the higher T_m, the more ordered the β-modification), the second one appears more sporadically, often during the first heating. It has been attributed either to (i) the melting of the original β_1-phase into a more stable

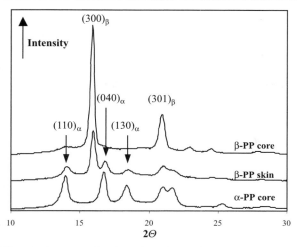

Fig. 1 WAXS diffraction pattern of a β-nucleated sample

β_2-structure during scanning [45, 46]; or (ii) to a perfection and thickening of existing lamellae without any change in geometry and order of the crystals [51]. Since the first heat scan accounts for the thermal history of the sample, the β-content is traditionally measured during the second heat in order to compare grades regardless of the way they were originally manufactured. According to our experience, if the melting enthalpy of the β-form contributes to 80% or more out of the total enthalpy, the nucleating agent can be considered as highly active and selective. The determination of the degree of crystallinity of both α- and β-phases is hazardous because the values of the melting enthalpies of both pure crystalline forms are not known with precision ranging from 146 to 240 J g^{-1} for $\Delta H(\alpha)$ [103–107] and from 113 to 195 J g^{-1} for $\Delta H(\beta)$ [103, 104, 108]. Based on a combination of classic WAXS measurements and original DSC configurations (e.g. cooling down of the nucleated molten PP sample to 100 °C only in order to avoid a β/α - recrystallization during the second heat [70, 109, 110]), Chen et al. proposed the most probable $\Delta H_{100\%}(\alpha)$ and $\Delta H_{100\%}(\beta)$ to be respectively 177 and 168.5 J g^{-1} [78]. A supplementary way to control the efficiency of a β-nucleation is to measure the crystallization temperature, T_c, during the cooling stage. The higher the difference in T_c between the non-nucleated and the corresponding β-nucleated grade is, the more developed is the β-phase. However, T_c is dependent on the cooling rate: the lower the cooling rate is, the higher is T_c. Recent data from Marco [27] showed T_c to evolve—for a 0.05 wt % NU-100 nucleated PP—following (with V_c, the cooling rate):

$$T_c = -4.62 \ln (V_c) + 135.7 \qquad (3)$$

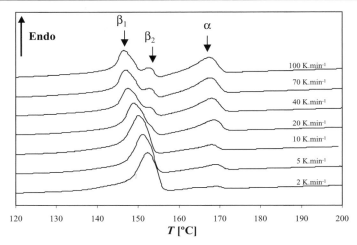

Fig. 2 DSC traces of a β-nucleated grade with cooling rates ranging from 2 to 100 K min^{-1} obtained in the second scan with a heating rate of 10 K min^{-1}. The melting point, T_m, of the β_1-modification is linearly correlated with V_c, the cooling rate, following: $T_m = -1.564 \ln(V_c) + 153.57 (R^2 = 0.9932)$

Optical microscopy offers an interesting possibility to characterize the β-phase, as a result of the differences in birefringence between the β- and α-crystals [93]. Its use will be highlighted in different sections of this manuscript. Although image analysis could provide some semi-quantitative information about the amount of the β-fraction, the optical micrographs are generally evaluated qualitatively by the eyes of a man-of-the-art as control for the WAXS/DSC data.

2.3
Assessment of Mechanical Performance

This section aims to summarize the main ways classically used to assess the mechanical performance of a material. For some of them, a more precise description will be given in the relevant sections.

Toughness represents the ability of a grade to dissipate energy during loading. In industrial routine, the determination of notched impact strengths is well established. It consists of measuring the energy to break of a standardized specimen under dynamic load conditions (from 1 to 4 m s^{-1}) with strikers of given nominal energy. They rely on international standardized procedures (e.g. ISO 179 for the Charpy test or ISO 180 for the Izod one). In sophisticated experimental setups, the displacement of the specimen during deformation can be recorded. From the load-displacement curves, it is possible to gain detailed information about the fracture behavior of the studied grades.

In Charpy-like tests, a severe triaxial stress state prevails at the sample crack tip. Softer modes of loading can be of interest for specific applications. Biaxial tests like impact falling weight are therefore often used. Plaques of defined dimensions are impacted with a hemispherically tipped dart of given diameter at selected test speed(s) (often 3 m s^{-1} and more).

Besides their inherent dynamic effects, one of the weaknesses of the previous mentioned tests is that they constitute single points of measurements: they are performed at one test speed and one temperature. To somewhat simulate the viscoelastic behavior of polymers, measurements over a wide range of temperatures or test speeds have to be carried out. The target of such experiments is to determine the brittle/ductile transition of the investigated materials: the lower the temperature (or the higher the test speed) at which the transition occurs, the tougher are the grades. Basically they can be assessed with any kind of loading provided that the testing machines are equipped with a climate chamber and/or with the possibility to vary the test speed. If the fracture history of the samples cannot be recorded the unique variable is the energy to break; if not, several relevant parameters can be followed (maximum load, energy to yield, ...). However, the fracture parameters used in this paper are not (or seldom) intrinsic (e.g. geometry independent) values and should be considered as toughness descriptors intended to rank materials.

Besides impact testing, quasi-static measurements are carried out to assess the Young modulus, E, the yield stress, σ_y, and the elongation at break, ε_{break}, as the most current parameters. They follow international standards (e.g. ISO 527 for tensile tests, ISO 178 for bending measurements).

Many of the studies use the melt flow rate (*MFR*) as an indirect indicator of the molecular weight, M_w, of PP and their blends. We will stick to this procedure. To facilitate the understanding of this review, a rough correlation between *MFR* and M_w is provided in Eq. 4. The molecular weights (in g mol^{-1}) of a model series of PP produced with the same (but specific) Ziegler–Natta 4th generation catalyst system and identical reactor settings have been measured by SEC (size exclusion chromatography) using a polystyrene calibration. Their *MFR* (230 °C/2.16 kg) ranging from 0.3 to 40 dg min^{-1} have been determined according to ASTM 1238 [111].

$$M_w = -155\,000 \ln(MFR) + 750\,000 \tag{4}$$

3
Toughness in β-Polypropylene

The mechanical performance of a polymer is governed by its intrinsic molecular architecture (e.g. molecular weight, tacticity, polydispersity, ...) and by extrinsic parameters such as the test speed, the test temperature, the stress state and the processing conditions. They will be reviewed for β-PP, a field for which numerous literature data are available. However, the proposed conclusions are globally valid for filled and rubber-modified β-PP.

3.1
Influence of Intrinsic Characteristics

3.1.1
Influence of Molecular Weight

The fracture resistance of a homopolymer is controlled by (i) the number of tie-molecules, which act as stress transducers between the crystallites; and (ii) by the disentanglement resistance of the individual chains [112–115]. The higher the molecular weight, the more likely both factors are high; β-nucleated polypropylene does not infringe this rule [33, 72, 74, 116].

To confirm it, PP resins with melt flow rates, MFR, from 0.3 to 40 dg min^{-1} have been produced [111]. Calcium pimelate, the proprietary technology of Borealis, has been used as a β-nucleating agent. The grades were considered to be fully nucleated, since their β-content, measured by DSC (2nd scan, 10 K min^{-1}), was about 80% ($\pm$ 2%). Their performance was compared with that of the corresponding non-nucleated materials, which exhibit an almost pure α-crystalline structure.

As obvious from Fig. 3a, the flowability of the resins has a strong influence on their impact behavior at room temperature assessed by their Charpy double edge-notched impact strengths (ISO 179/1fA). As expected, for both non-nucleated and β-nucleated series, a higher molecular weight favors the resistance to crack propagation due to the high density of inter- and in-

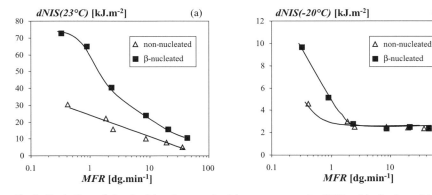

Fig. 3 Evolution of the double edge-notched impact strength, $dNIS$, with the logarithm of the melt flow rate, MFR, **a** at room temperature and **b** – 20 °C. The test speed was about 3.8 m s^{-1}, the specimens were injection molded

tralamellar links which promotes shear deformation and thus matrix flow. Moreover, the β-PPs perform significantly better than their non-nucleated counterparts over the whole range of MFR at 23 °C. At low temperatures (Fig. 3b), this improvement is less pronounced and disappears for melt flow rates higher than 2 dg min^{-1}. The ratios between the fracture energies of the β-nucleated and non-nucleated grades at defined MFR are given in Fig. 4. They decrease with increasing MFR, slightly at room temperature, more sharply at – 20 °C. However, it cannot be concluded from these values that

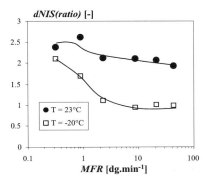

Fig. 4 Ratios between the double notched impact strengths of β-nucleated and non-nucleated grades in between melt flow rates, MFR, of 0.3 to 40 dg min^{-1} at room temperature and – 20 °C. The test speed was about 3.8 m s^{-1}, the specimens were injection molded

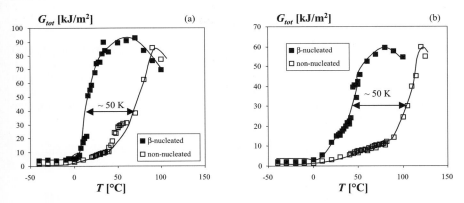

Fig. 5 Evolution of the fracture energy, G_{tot}, with the temperature, T, for non-nucleated and β-nucleated resins with different flowabilities: **a** MFR: 0.3 dg min^{-1} and **b** MFR: 2 dg min^{-1}. The ductile-brittle transition temperature was chosen in a somewhat arbitrary manner as the temperature corresponding to half of the maximum of G_{tot} in the considered MFR range. It reflects the transition from a semi-ductile to a fully ductile behavior, without breaking of the tested specimen. The test speed was about 1.5 m s^{-1}, the specimens were injection molded

β-nucleation is less effective with a low molecular grade than with a high molecular one. The temperatures at which the ductile-brittle transitions, T_{db}, occur for selected non-nucleated and β-nucleated sets support our conclusion (Fig. 5): β-nucleation lowers the T_{db} by about 50 °C independently of the flowability of the neat polymer. Although these tests were performed with single notched specimens, they can be correlated to a certain extent to those carried out on double-edge notched specimens. Sect. 3.2.4 deals in more detail with the influence of the local stress concentration on the mechanical responses of β-iPP.

3.1.2
Influence of Polydispersity

The molecular weight distribution (MWD) of a PP can be controlled either during polymerization or in a compounding step. While high polydispersity (broad MWD) is achieved in the reactor by polymer design, a narrow M_w/M_n distribution can be obtained by peroxide visbreaking (which induces a concomitant reduction of M_w). The use of selective catalysts allows both possibilities.

Studies dealing with the influence of MWD on the toughness of β-PP are sparse. A metallocene β-PP, the archetype of a material with a narrow polydispersity (PI: 2–3), was recently shown to exhibit a lower relative increase of the fracture resistance (between its non-nucleated and β-modified versions) than its Ziegler–Natta (ZN) homologue (PI: 5–6) [117]. This result was attributed to the worse crystallization ability of a single-site grade—resulting in

Table 1 Influence of visbreaking on the fracture toughness (NIS according to 179/1eA) and stiffness measured in bending after ISO 178, $E_{bending}$, of non-nucleated and β-nucleated grades. CR means controlled rheology, PI polydispersity. $\Delta H(\alpha)$ is the melt enthalpy of the α-phase (2nd scan), $\Delta H(\beta)$ that of the β-modification (2nd scan), T_c is the crystallization temperature determined by DSC with the scanning program: $+10/-10/+10$ K min^{-1}

MFR final [dg min^{-1}]	MFR initial [dg min^{-1}]	Visbreaking [yes/no]	Polydispersity [–]	Nucleation [no/β]	NIS(23 °C) [kJ m^{-2}]	$E_{bending}$ [–]
8.5	8.1	no	5.9	no	3.1 ± 0.1	1435
8.5	0.2	yes	3.4	no	4.1 ± 0.3	1220
8.4	0.7	yes	3.4	no	3.6 ± 0.2	1265
8.6	8.1	no	5.9	β	7.3 ± 0.2	1285
7.9	0.2	yes	3.4	β	7.8 ± 0.2	1110
8.1	0.7	yes	3.4	β	7.3 ± 0.2	1180

a two times higher crystallization half-time in between 115 and 130 °C—and correlated to its enhanced amount of chain folding irregularities [118].

Table 1 summarizes our own conclusions for controlled rheology (e.g. visbroken) grades. The benefits of visbreaking β-PP are limited and dependent on the degradation length. Two competitive effects seem to govern the impact resistance of β-modified products: their narrower M_w/M_n, known to improve the toughness (as confirmed in Table 1 for non-nucleated grades), and their somewhat lower crystallization temperature, T_c, and crystallinity (assessed by the melt enthalpy of the main DSC peak) compared to the reactor grade, leading to reduced mechanics. As a result, because visbreaking has a detrimental effect on stiffness, peroxide degradation of β-nucleating grades makes little sense.

3.1.3
Influence of a Matrix with Low Ethylene Content

Random E/P copolymers consist of PP chains in which small amounts of ethylene (C_2) are more or less randomly distributed. The C_2 units act as defects for the regularity of chain configuration and reduce the overall crystallinity of the polymer.

The consequence of this matrix disturbance is nicely illustrated in Fig. 6 for grades with a melt flow rate of about 1 g/10 min containing 0, 4.6 and 8.3 mol % randomly distributed ethylene. The development of the β-phase is progressively slowed down, before being stopped with an increase in the fraction of comonomer; at 8.3 mol %, almost only α-crystals are present in the system (Table 2). Although the fracture resistance is positively correlated with the ethylene content of the β-nucleated resins, a weaker relative impact en-

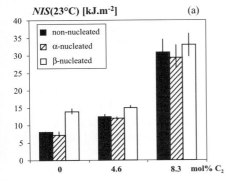 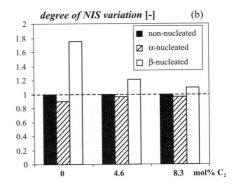

Fig. 6 Notched Impact Strengths (*NIS*) of the non-nucleated, α-nucleated and β-nucleated resins investigated in Sect. 3.1.3; **a** absolute values; **b** relative values, the non-nucleated grade having been taken as reference for each C_2-content. The test speed was about 3.8 m s^{-1}, the specimens were injection molded

Table 2 Main characteristics of the investigated resins of Sect. 3.1.3. All data have been measured on DSC scans with a program of + 10/ – 10/ + 10 K, T_c is the crystallization temperature, $\Delta H(\alpha)$ and $\Delta H(\beta)$ the enthalpy of fusion of resp. the α and β-forms (2nd scan)

Material	Nucleation $[no/\alpha/\beta]$	C_2-content [mol]	MFR [dg min^{-1}]	T_c [–]	$\Delta H(\alpha)$ [J g^{-1}]	$\Delta H(\beta)$ [J g^{-1}]	β-Content %
N-0	no	0	2	111.5	96	0	0
α-0	α	0	2.1	120.9	102.5	0	0
β-0	β	0	2.3	122.5	18.4	100.4	85
N-4	no	4.6	2.1	104.1	88.7	0	0
α-4	α	4.6	2.2	110.1	87.8	0	0
β-4	β	4.6	2.1	108	27.3	69.5	72
N-8	no	8.3	2	95.5	62.4	0	0
α-8	α	8.3	2.2	96.7	69.8	0	0
β-8	β	8.3	2.1	96.2	65.4	0.4	1

hancement is observed with C_2-rich random E/P copolymers when compared to their corresponding non-nucleated and α-nucleated counterparts. Hence a toughness improvement of 75% is reached with β-nucleating a neat homopolymer; only 20% is gained with the copolymer with 4.6 mol % of C_2 and nothing with the one with 8.3 mol % of C_2 (considering the standard deviation of the measurements).

These observations are in accordance with those of Fujiyama [38], Zhang [46, 119], Varga [10, 51, 56] and others [78, 120]. They might be explained by a combination of thermodynamic and kinetic effects. On the one hand, a preferential development of α-crystals arises from the lower growth rate of the β-phase (G_β) compared to that of the α-phase (G_α) over a wide range of temperatures in random E/P copolymers [51, 70, 110, 121], the crossover temperature for which $G_\beta > G_\alpha$ being shifted to lower temperatures with increasing ethylene content [70]. On the other hand, a thermodynamically controlled transition from β- to α-crystallization has been proposed to lead to the formation of α-lamellae on the surface of growing β-spherulites resulting in $\alpha\beta$-twin-spherulites during cooling [122, 123]. This phenomenon might be lowered by increasing the density of the β-nuclei in the resin, e.g. by increasing the concentration of the nucleating agent, as remarked correctly by Varga [51].

3.2
Influence of Extrinsic Parameters

3.2.1
Influence of the Content of Nucleating Agent

The forced development of the β-PP modification by addition of selective nucleating agents is a heterogeneous nucleation process. The growth of the β-phase is governed by the number of nuclei present in the system. Increasing the content of the nucleating agent can therefore be thought to promote its efficiency in terms of preferential development of β-lamellae and subsequent improvement of fracture resistance up to a saturation level.

This expectation was confirmed using calcium pimelate as the nucleating agent in a concentration range up to 3000 ppm. The evolution of the microstructure as observed in polarized light microscopy with increasing amounts of β-promoter is given in Fig. 7. The 50 μm large α-spherulites disappear progressively, leading to a finer and finer β-superstructure characterized by spherulites of about 1 to 5 μm. Fig. 8 highlights the evolution of the amount of β-phase, the crystallization temperature and the mechanical performances (fracture resistance at 23 °C and stiffness) as a function of the amount of calcium pimelate. The base resin is fully nucleated at a concentration of 500 ppm of calcium pimelate as obvious from Fig. 8 from the β-PP amount (> 80%) and by the crystallization temperature assessed by DSC (cooling program: + 10/ – 10/ + 10 K min^{-1}). Above this concentration, this nucleation level could be maintained as shown by the plateau values exhibited by both parameters. Compared to the neat base polymer, this almost complete β-nucleation is accompanied by high fracture resistance values (+ 800%) and by a decrease of 200 MPa of the flexural stiffness (– 15%). However, with other nucleating agents the impact toughness of β-PP goes through a marked maximum with increasing the amount of the β-promoter. This is actually the case for a concentration of about 0.01% of γ-quinacridone [21, 32, 36, 124] and around 0.03% of N,N-dicyclohexyl-2,6-naphthalene dicarboxamide [63, 64, 68, 84, 125] as described in Fig. 9 and 10 and confirmed by various groups. This phenomenon could have been partly foreseen by WAXS (Fig. 9), but neither by DSC (Fig. 10) nor by the crystallization temperature of the β-phase (Figs. 9 and 10). Sterzynski has suggested that this maximum results from a synergy between a higher germination density and the respective growth kinetics of the α and β crystal structures [42]. Recent convincing experiments based on NU-100 as the nucleating agent have correlated it with a minimum in the lateral size of the β-crystallites and with a maximum in the long-period of the β-phase [64]. The key role of the amorphous phase in the deformation process and thus on the impact toughness has been pointed out. Since the size of the amorphous phase is positively correlated with that of the

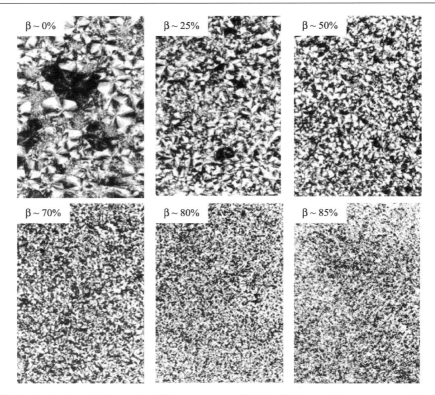

Fig. 7 Evolution of the spherulitic structure of PP with increasing amounts of calcium pimelate (from 0 to 2000 ppm) leading to increasing amounts of β-modification as observed by optical microscopy. Micrographs (425 × 650 µm) kindly provided by J. Wolfschwenger

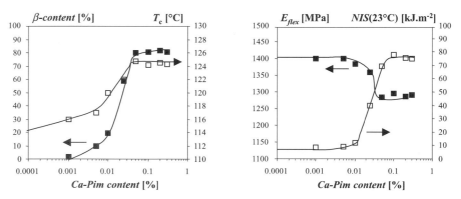

Fig. 8 Evolution of the amount of β-phase (DSC), the crystallization temperature, T_c, and the mechanical performances (fracture resistance at 23 °C and stiffness) as a function of the amount of calcium pimelate

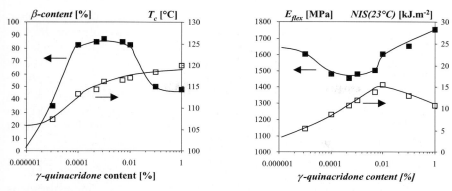

Fig. 9 Evolution of the amount of β-phase (WAXS), the crystallization temperature, T_c, and the mechanical performances (fracture resistance at 23 °C and stiffness) as a function of the amount of γ-quinacridone. Data adapted from Fujiyama [21]

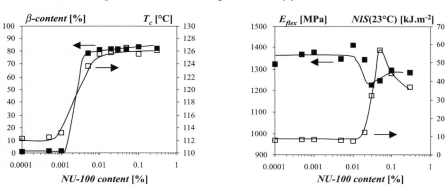

Fig. 10 Evolution of the amount of β-phase (DSC), the crystallization temperature, T_c, and the mechanical performances (fracture resistance at 23 °C and stiffness) as a function of the amount of N,N-dicyclohexyl-2,6-naphthalene dicarboxamide (NU-100)

long period, the maximum in toughness has been attributed to easier matrix flow arising from the increased molecular mobility of the system.

3.2.2
Influence of the Type of Nucleating Agent

Under standard processing conditions, PP crystallizes in between 5 and nearly 100% in the β-modification depending on the nature of the β-nucleating agent [13–31, 126, 127]. Industrially, only highly selective β-promoters are of interest since the toughness is positively correlated with the β-content in a more or less straightforward way (see Sect. 3.2.2).

The interconnection toughness/β-content has been nicely illustrated by Tordjeman et al. [76]. A special sample preparation procedure made it possible to vary the amount of β-phase without changing any other parameter, especially not the size of the spherulites (which is known to influence dramatically the fracture resistance). Essential Work of Fracture (EWF), a fracture mechanics method for ductile mode of deformations, was used for data reduction. The near plane-stress essential work of fracture, w_e (expressed in J m^{-2}) accounting for the energy needed to create two new surfaces during loading, was shown to evolve linearly with the amount of β-phase in the sample. For three-point bending tests performed with 3 mm thick specimens at 500 mm min^{-1}, it can be described by:

$$w_e \approx 65.2\,K_\beta + 999 \tag{5}$$

where K_β is the amount of β-phase in % measured by DSC. Under the same test conditions, the non-essential work of fracture, Cw_p (in J m^{-3}), accounting for the volumetric plastic energy associated with the damage process, correlates also linearly with x_β following:

$$Cw_p \approx 1.5\,K_\beta + 272 \tag{6}$$

Moreover, other more classical mechanical descriptors like yield stress, Young modulus or elongation at break evolved also monotonically with K_β.

Historically, γ-quinacridone has been the first nucleating agent used. Its activity at lowest concentrations made it extremely attractive. However, alternatives had to be found since the induced red coloration was penalizing for numerous applications. Fujiyama has used quinacridonequinone [21], which gave a slight yellow coloration; Borealis has developed its colorless nucleating agents [19] as well as Atofina [25] or New Japan Chemicals with the commercial N,N-dicyclohexyl-2,6-naphthalene dicarboxamide (e.g. NU-100) [17].

Systematic studies about their relative mechanical performance are rare. The pioneering work of Fujiyama who compared γ-quinacridone with quinacridonequinone shows in an incontestable way the non-equivalence of different highly efficient β-promoters [21]. For an optimal concentration, the β-nucleated resins differed in their notched impact strength, yield stress, elongation at break and elastic modulus. Our own investigations confirmed this feature.

A series of iPP with different flowabilities was tested. They were nucleated with 0.1% calcium pimelate (Ca-Pim), 0.1% NU-100 or 0.002% γ-quinacridone. The β-content of the materials was measured by DSC (2nd scan, 10 K min^{-1}); it was not dependant on the MFR of the resins and was about 80% for Ca-Pim, 86% for NU-100 and 84% for γ-quinacridone, suggesting the resins to be fully nucleated. The results for the notched impact toughness tested at 23 °C (NIS 179/1eA) are reported on Fig. 11a (absolute values) and in Fig. 11b (relative values, normalized by the NIS of the non-nucleated iPP). Except for NU-100 above a melt flow rate of 15 g/10 min, β-nucleation increased the room temperature toughness of PP homopolymers. The following conclusions could be drawn:

- β-nucleation promotes energy dissipation in a range of MFR from 0.2 to 40 g/10 min;
- the apparent efficiency of β-nucleation depends on the MFR: the higher the MFR is, the lower is the *apparent* efficiency of β-nucleation;
- the three nucleating agents exhibit specific field and/or level of performance: while calcium pimelate efficiently promote toughness over the whole range of MFR, NU-100 was rather adapted for low flowable resins (up to an

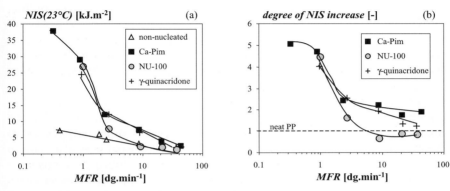

Fig. 11 a Notched impact strengths, *NIS*, at room temperature for non-nucleated, calcium pimelate nucleated, NU-100 nucleated and γ-quinacridone-modified PP homopolymers over a wide range of MFR; **b** degree of fracture resistance improvement with the different nucleating agents—the values are normalized with the *NIS* of the non-nucleated corresponding neat resin taken to evolve as follows: $NIS(\text{non-nucleated}) = -1.26 \ln(MFR) + 6.01$. The test speed was about 3.8 m s^{-1}, the specimens were injection molded

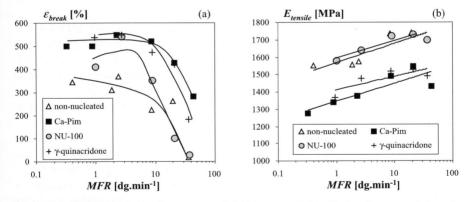

Fig. 12 a Elongation at break, $\varepsilon_{\text{break}}$, and **b** Young modulus, *E*, for the non-nucleated, calcium pimelate nucleated, NU-100 nucleated and γ-quinacridone-modified PP homopolymers over the investigated MFR range. The test speed was 50 mm min^{-1} for the determination of $\varepsilon_{\text{break}}$, 5 mm min^{-1} for the evaluation of *E*. The specimens were injection molded dog bones

MFR of 2 dg min^{-1}) and γ-quinacridone exhibits an intermediate behavior.

These conclusions were consolidated with tensile tests performed at moderate strain rates (Fig. 12, $v = 5$ mm min^{-1}, $d\varepsilon/dt \sim 1$ s^{-1}). The elongation at break, ε_{break}, is an indirect indicator for energy dissipation [Fig. 12(i)]. As expected, by analogy with the impact tests, ε_{break} decreased with increasing MFR. The lower efficiency of NU-100 in the higher MFR range was obvious: the resins became as brittle as the non-nucleated homopolymers. This might be linked to the stiffness of the resins, which was at the same level as the corresponding neat PP (Fig. 12b).

3.2.3
Influence of Processing Conditions

The development of the β-modification is controlled by the relative crystallization thermodynamics and kinetics of the stable α-modification and of the smectic phase towards the metastable β-phase. For PP homopolymers, it is generally accepted that under isothermal conditions, the α-phase grows more rapidly at temperatures below 105 and above 140 °C than its counterpart, which in turn is more prone to develop in between these two temperatures in the presence of selective β-promoters [52, 70, 122]. An elegant way to get fully nucleated β-PP specimens would consist of pressing β-PP pellets above their melting temperature (ideally more than 250 °C to erase any α-nuclei in the system), cool the melt quickly up to a crystallization temperature in-between 100 and 130 °C, let the sample crystallize, and then quench it to room temperature [70]. However, such a processing method is too time-consuming to be of industrial relevance.

Much of the experimental data dealing with β-PP films, fibers, pipes, compression and injection molded specimens might be explained by considering (i) the cooling rate; (ii) the shear rate which develops during processing; and (iii) the thickness of the end-product. Indeed, due to the poor heat transfer in PP, the temperature gradient within the sample and/or between the sample and its environment drives the development of any PP-structure to a large extent.

In thin specimens (e.g. films and fibers to oversimplify), moderate cooling rates are preferred: they promote the development of the β-phase in temperature ranges where β-PP competes with the smectic/mesomorphic phase (high cooling rates/quenching) or with the α-modification (very low cooling rates). This can be achieved by increasing the temperature of the cooling medium in which the polymer is immersed after processing (air/water) [34, 128, 129], by increasing the fiber diameter [128, 129], by increasing the cast/chill roll temperature (up to the optimum temperature to get β-PP in a pure form, e.g. 100–110 °C) [130] and by monitoring the extrusion/spinning temperature [128, 131, 132]. In the case of fibers, their stabilization temperature (in between 110 and 160 °C) and draw ratio should be kept as low as possible within the frame of the application to limit a β/α heat-induced transformation [128, 133, 134]. Moreover, because of the limited formation of shear-induced α-nuclei, low extrusion/spinning/take-up speeds have been shown to favor the growth of β-PP in general [70, 129, 132–135], with the exception of that shown in [128]. They are, however, detrimental for the throughput.

In thicker specimens, the development of α-spherulites at temperatures above 130 °C can be avoided by setting faster cooling rates. Once the nuclei have germinated, the α-modification is, indeed, thought to grow further even if the local temperature would be favorable to the development of the β-phase [70, 122]. This is especially the case for compression molded samples where the molecular orientation is marginal and has a negligible influence on the crystallization process. The influence of the cooling rate on the impact toughness of γ-quinacridone nucleated PP is exemplified in Fig. 13. Four millimeter thick samples were pressed at 250 °C; they were cooled (i) slowly (by allowing the specimens to cool down in the mold, $V_c \sim 2\,\mathrm{K\,min^{-1}}$); (ii) in a "conventional" way at $10\,\mathrm{K\,min^{-1}}$; and (iii) rapidly (by cooling the mold with cold water, $V_c \sim 20\text{--}30\,\mathrm{K\,min^{-1}}$). The mechanical performance was assessed by determination of the ductile-brittle transition at constant test speed ($\sim 1.5\,\mathrm{m\,s^{-1}}$) of notched samples over a broad temperature range using an instrumented Charpy pendulum. A gap of 40 K could be recorded between both cooling extrema: while the fastest cooling rate lead to a ductile-brittle transition temperature, T_{db}, of about 6 °C, the slowest exhibit a T_{db} around 40–50 °C. This feature could be correlated with the micromorphology of the materials: excellent impact performance was associated with the smallest spherulitic size (Fig. 14). For injection molded grades, things be-

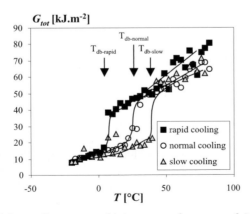

Fig. 13 Influence of the cooling rate on the impact performance of β-modified PP (MFR: $0.3\,\mathrm{dg\,min^{-1}}$). The specimens were compression molded and tested at $1.5\,\mathrm{m\,s^{-1}}$. The arrows indicate the temperature at which the ductile-brittle transition occurred

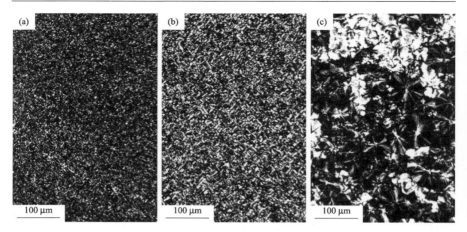

Fig. 14 Spherulitic structure of compression molded samples (core) obtained with different cooling rates: **a** rapid, **b** conventional and **c** slow

come even more complicated. In general, the observed superstructure is smaller than for the corresponding compression molded grades because this latter is associated with a less rapid quenching. As a consequence, injection molded β-PP is, as expected, more resistant to crack propagation than its pressed counterpart. Data sets from Nezbedova [63] and Fujiyama [21] support this assertion. However, flow and thermal conditions in the mold which lead to the widely described, but complex, skin-core morphology make any analysis difficult [32, 33, 38–40, 42, 53, 72, 74, 77, 109, 136–138]. This is especially due to the gradient of β-phase which develops in the sample: the core, submitted to intense shearing, contains less β-PP than the core, where the crystallization has been proposed to occur in a quiescent melt [48, 53, 72, 77, 124, 137, 138]. As a result, any method which reduces the size of the β-poor skin is welcome. Different ways to promote the development of the β-phase have been identified: increasing the injection speed [63, 109, 137, 138], optimizing the melt temperature [32, 38] and increasing the mold temperature [36, 42, 63, 74, 84, 139].

3.2.4
Influence of Stress-State

The stress-state has been widely recognized to play a key-role in the magnitude of the developed fracture resistance. As far as we know, only Karger-Kocsis dealt with this topic, despite its practical implication. He compared a unixial in-plane load (instrumented tensile impact, $3.7\,\mathrm{m\,s^{-1}}$) with a biaxial out-of-plane sollicitation (impact falling weight, $10\,\mathrm{m\,s^{-1}}$) using homopolymers with different flowabilities (MFR: 0.8, 5.5 and $13\,\mathrm{dg\,min^{-1}}$) [74]. The energies to break values, G_{tot}, obtained with both methods are summarized in

Table 3. The toughness improvement induced by the β-modification, rationalized in terms of the ratio $G_{tot}(\beta\text{-nucleated})/G_{tot}$ (non-nucleated), is markedly stronger for the series tested under biaxial conditions. As both sets of investigated samples were injection molded, the skin-core structure (discussed in Sect. 3.2.3) could be inferred to be at least partly responsible for this feature. This was, however, confirmed in a recent work, where impacted β-PP samples (tested at 5 m s^{-1}, $T = 23$ °C and -20 °C) with or without skin layers showed the same fracture performance within the limits of the experimental error [116].

The influence of the stress state can be further appreciated by considering the mechanical response of a non-nucleated and β-modified low flowable PP assessed by (i) conventional notched; (ii) double-edge notched; and (iii) unnotched Charpy measurements (Table 4). While the latter were apparently insensitive to the presence of any β-phase, an increased severity of the tests allows the β-modification to show its full potential. Compared to a severe notched Charpy test, where a triaxial stress state prevails in the crack tip, impact falling weight tests are smoother, even at high loading rates. Although being both submitted to a dynamic bending mode of deformation, they differ (i) in the extent of their elementary tensile and compression stress contribu-

Table 3 Energy to break, G_{tot}, and energy ratios, $G_{tot}(\beta\text{-nucleated})/G_{tot}$(non-nucleated), obtained with instrumented tensile impact and impact falling weight test using non-nucleated and β-modified PP homopolymers exhibiting an MFR of 0.8, 5.5 and 13.4 dg min^{-1}. Data taken from Karger-Kocsis [74]

MFR	Instrumented tensile impact			Impact falling weight		
	G_{tot} (non-nucl.)	G_{tot} (β-nucl.)	Ratio (β/non)	G_{tot} (non-nucl.)	G_{tot} (β-nucl.)	Ratio (β/non)
[dg min^{-1}]	[kJ/m^2]	[kJ/m^2]	[–]	[J/mm]	[J/mm]	[–]
0.8	375	521	1.4	2.3	6.7	2.9
5.5	104	281	2.7	0.7	4.4	6.3
13.4	100	191	1.9	0.7	1.4	2.0

Table 4 Mechanical performance of a non-nucleated and β-nucleated PP with an MFR of 0.3 dg min^{-1}. Comparison of notched, double-edge notched and unnotched Charpy tests

	G_{tot}(non-nucl.) [kJ/m^2]	G_{tot}(β-nucl.) [kJ/m^2]	Ratio (β/non) [–]
Notched	7.2	37.8	5.2
Double-edge notched	30.5	72.8	2.6
Unnotched	23.5	19.3	0.8

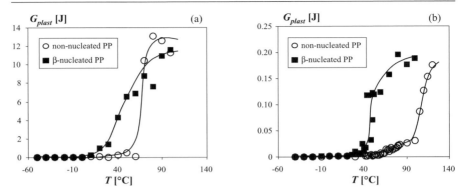

Fig. 15 Evolution of the plastic energy, G_{plast}, of a non-nucleated and of a β-nucleated PP in between – 60 and 120 °C: **a** puncture tests, **b** Charpy tests. Tests performed on injection molded specimens

tions; (ii) in the localization of their weak zone (highly localized in notched Charpy samples/rather distributed in the unnotched puncture plaques); and (iii) in the influence of the skin layer on the mechanical properties. Fig. 15 shows a direct comparison between the ductility of a non-nucleated and of a β-modified PP (MFR: 2 dg min^{-1}) measured on (i) 2 mm thick impacted injection molded plaques at 4.4 m s^{-1} (Fig. 15a) and (ii) 4 mm thick-notched Charpy specimens at 1.5 m s^{-1} (Fig. 15b) in between – 60 and 120 °C. The ductility was assessed by the plastic energy, e.g. the difference between the energy to break and the peak energy (i.e. the energy up to the maximum of the force, often called initiation energy); the higher this plastic energy, the less brittle is the grade. As expected, β-nucleation improves the fracture resistance of the investigated homopolymer. But, this enhancement is dependent on the testing procedure: whereas a shift of about 50 °C could be recorded with the Charpy testing procedure, only 20 °C could be gained with the perforation measurements.

3.2.5
Influence of Test Speed and Temperature

As the superior fracture resistance of β-PP towards its non-nucleated version has been largely consolidated using standard mechanical characterizations, the influence of the temperature or test speed on its mechanical performance has received more attention over the last few years [48, 50, 55, 72, 77, 78, 119, 140]. Tjong et al. reported that the Charpy impact energy at 20 °C of a β-PP (MFR: 4 dg min^{-1}) decreased from 12 to 6 kJ m^{-2} in between 0.76 and 4.8 m s^{-1}, while it varied from 6.5 to 4 kJ m^{-2} for its non-nucleated homologue [50]. In another paper, a similar β-PP exhibited 2 to 3 times the elongation at break of its counterpart in between 1 and 200 mm min^{-1}, be-

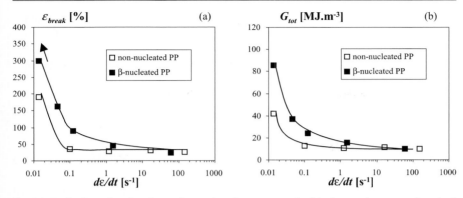

Fig. 16 Evolution of **a** the elongation at break, ε_{break}, and of **b** the total energy absorbed before fracture, G_{tot}, of β-modified and non-nucleated PP (MFR: 12 dg min^{-1})

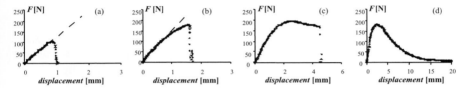

Fig. 17 Force-displacement curves corresponding to elementary material's behaviors: **a** brittle, **b** semi-brittle, **c** semi-ductile and **d** ductile

fore showing no more difference at 500 mm min^{-1} [48]. Similar trends were observed in between 0.001 and 1 m s^{-1} in our own study performed on MFR 12 grades (described in [77]) for both ε_{break} and G_{tot}, the total energy absorbed before fracture (e.g. the integral $\int \sigma d\varepsilon/dt$ up to fracture) as shown in Fig. 16. Moreover, the yield stress, σ_y increases in a logarithmic way with the strain rate, like any other polymer [48, 55, 77, 119, 141–144], with plastic flow becoming less favored at higher test speeds.

The effect of temperature on the mechanical response of β-nucleated grades was investigated by several groups [72, 77, 78, 140]. In general β-PP was superior to the non-nucleated reference, although the difference between them was less pronounced with decreasing temperature. To get a deeper insight into this topic, it might be worthwhile to introduce the four successive fracture behaviors and the associated predominant deformation mechanisms that a material exhibits. With decreasing speed (or increasing temperature), one can observe (Fig. 17):

- A brittle behavior. The force-displacement (F-d) curves are linear elastic. The crack propagates in an unstable way. The damage mechanisms have not been initiated before failure. The fracture surfaces are very smooth and mirror-like. At the microscopic scale, they are believed to be associated with the development of a single crack.

- A semi-brittle behavior with pronounced non-linearity prior to unstable crack propagation. Fracture occurs before yield stress. The fracture surfaces are rough, probably composed of several planes of macrocracks, initiating within the crazes at different times and slightly different planes. The size of the rough whitened zone decreases with increasing test speed, implying a decrease in the extent of multiple crazing. This whitening accounts for changes in the refractive index of the material as a result of voiding in form of matrix crazing or particle cavitation, both of them being limited in the case of a semi-ductile behavior.
- A semi-ductile behavior with initiation and partial development of the damage mechanisms before unstable fracture. Near the crack tip, the fracture surface is fully whitened accounting for the limited stable growth of the crack (the extent of whitening decreased with increasing test speed); far from the crack tip, it is rough like in the semi-brittle case. A mixed mode between small scale yielding and multiple crazing is believed to be associated with this mode of fracture.
- A ductile behavior accounting for intense shear yielding. It is characterized by stable crack propagation, entire stress-whitening of the fracture surface and is accompanied by more or less pronounced shear lips.

Karger-Kocsis recorded the different fracture behaviors of non-nucleated and β-modified PP (MFR: 0.8 dg min^{-1}) tested in a three-point bending configuration at 1 m s^{-1}; at 23 °C, α-PP was semi-ductile and β-PP ductile with a plastic hinge; at -40 °C α-PP was brittle, β-PP ductile [72]. The descriptors from the linear elastic fracture mechanics (LEFM), K_C, the stress intensity factor, and G_C, the energy release rate, used to quantify the toughness correlated well with the fracture picture. This conclusion is also valid for

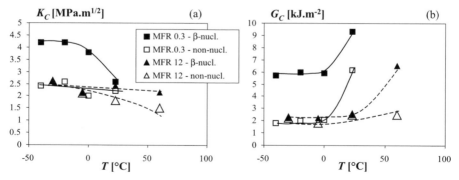

Fig. 18 Evolution of the **a** stress intensity factor, K_C, and the **b** energy release rate, G_C, for a low flowable and a medium flowable non-nucleated and β-modified PP. The legend is the same for both parts of the figure. Injection molded specimens tested at 1 m s^{-1} in three-point bending [78] and compact tension [77] configuration

PP with different flowablities, extracted from [77, 78], depicted in Fig. 18. Generally, with increasing temperature, K_C decreases—reflecting its relation to the yield stress, especially when the grades are not brittle and/or not tested in plane strain conditions—and G_C increases as expected from the progressively enhanced molecular mobility of the materials. The stronger improvement above refrigerator temperatures correlates undeniably with the glass transition temperature; T_g, of PP (T_g is in between – 10 and 10 °C depending on the PP characteristics and on the T_g determination method). For the PP with MFR 12 dg min^{-1}, this spectacular enhancement is obviously shifted to $T > 40$ °C and reserved to the β-modified resin, accounting for the intrinsic lower toughness of high(er) flowable materials, which becomes particularly visible under impact conditions. This feature might all the more be accentuated by the nature of the test used (three-point bending for the low MFR grades, compact tension Mode I for the MFR 12 materials).

Figures 19 and 20 summarize the fracture behavior of the previously mentioned PP with MFR 12 dg min^{-1} in between – 30 and 60 °C. Details concerning the experimental procedure, the F-d curves and the data reduction according to the principles of the LEFM are given in [77]. The aim of this section is to correlate the relative capacity of both systems to absorb input energy up to a deformation corresponding to F_{max} (G_{ini}) and up to fracture (G_{tot}; Fig. 19) with their deformation maps deduced from their Fd-curves and careful observation of their fracture surfaces (Fig. 20). To take implicitly into account variations in specimen stiffness, arising from variations in temperature and rate dependence of the modulus; the test rate was sometimes expressed in terms of the crack tip loading rate, dK/dt, given by:

$$dK/dt = f(a/W) \frac{dF/dt}{B\sqrt{W}} \quad (7)$$

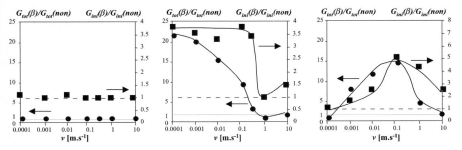

Fig. 19 The ratio between the initiation energy of a β-nucleated iPP (MFR: 12 dg min^{-1}) and its non-nucleated homologue [G$_{ini}$(β)/G$_{ini}$(non)] and the ratio between the energy to break [G$_{tot}$(β)/G$_{tot}$(non)]of the same grades over a wide range of test speeds at (from left to right) – 30 °C (same features than – 5 °C), 23 °C and 60 °C

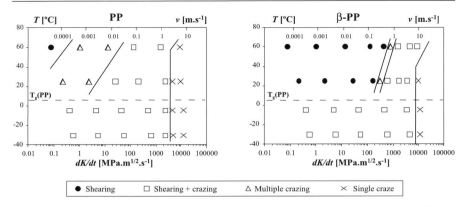

Fig. 20 Deformation maps of **a** a non-nucleated *i*PP and **b** its β-nucleated homologue (MFR: 12 dg min^{-1}) for different temperatures and crack tip loading rates as deduced from the fracture surfaces of compact tension specimens. A rough indication of the test speed is provided by the upper scale. Same grades as in Fig. 19

where B is the thickness of the specimen, W its width, dF/dt the initial slope of the force-time curve and $f(a/W)$ a tabulated dimensionless geometrical factor.

For $T > T_g$, the fracture behavior is sensitive to the crystalline structure. Whereas the β-nucleated grade is ductile up to 200 MPa m$^{1/2}$ s^{-1} (0.1 m/s) at room temperature and up to 800 MPa m$^{1/2}$ s^{-1} (0.4 m/s) at 60 °C, the non-nucleated homopolymer is relatively brittle over nearly the whole range of test speeds and temperatures (the exception being 0.0001 m/s at 60 °C). As a result, the ductile-brittle transition is shifted by about 3 decades of test speed thanks to the β-crystalline structure. To put this into perspective, it should be noted that such a shift corresponds approximately to an addition of 15 wt % of elastomer into a PP homopolymer. In the ductile range of β-*i*PP, its relative initiation energy (resp. energy to break) is 2 to 4 times (resp. 10 to 20 times) higher than those of its non-nucleated homologue. When β-PP becomes brittle, its behavior is nearly equal to that of the neat PP. For $T < T_g$ (where both materials are brittle over the whole investigated range of test speeds), no significant difference can be observed between the β-nucleated grade and its non-nucleated homologue.

4
Toughness of Filled β-Polypropylene

The combined modification of PP with fillers and β-nucleating agents has been seen as attractive for its potential to achieve a breakthrough in the somewhat frozen stiffness/impact balance of PP. It has been speculated that the loss in stiffness induced by the β-modification could be largely compensated for by the addition of fillers while retaining at least part of the exceptional fracture performance of β-PP. A synergy could actually only be found with inactive fillers like calcium carbonate (CaCO$_3$) [53, 54, 71, 75, 79, 84, 140, 145, 146], wollastonite [147], mica [40], glass flake [40] or glass fibers [148–151], which behave like inert components in the PP melt during the crystallization step, not with active fillers like talc [40, 71, 145, 152] which are strong promoters of the α-modification of PP. However, as most of the studies are based on single point measurements (performed at one speed and one temperature), the positive effect of inactive fillers was not systematically observed.

The typical impact/stiffness picture of a β-nucleated PP composite with inert fillers is given in Fig. 21 for CaCO$_3$ [53]. The parallel evolution of the E-moduli and notched impact strengths with the filler content constitutes a proof for the non-reactivity of chalk with PP. To reach a targeted toughness, it appears to be preferable to introduce more CaCO$_3$ in a β-nucleated system than producing a non-nucleated matrix with less filler. Fig. 22 exemplifies the case of talc, belonging to the active filler category [40]. With an increase in the filler content the difference in toughness between both non-nucleated and β-nucleated grades become weaker to insignificant above 20 wt % of talc. These features have been rationalized in terms of crystallization behavior in the melt. With talc (and on the contrary to inactive fillers), heterogeneous nuclei are formed at high temperature above the criti-

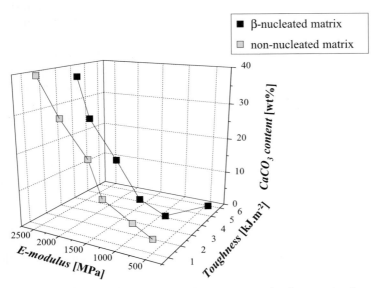

Fig. 21 Evolution of the modulus and notched impact strength of composites based on PP (MFR: 4 dg min^{-1}) and increasing amounts of calcium carbonate, CaCO$_3$. Data taken from Tjong [53]

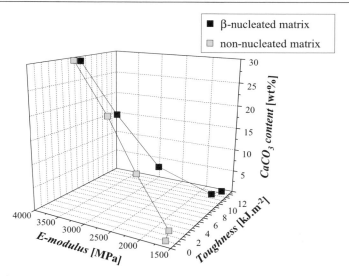

Fig. 22 Evolution of the E-modulus and Izod notched impact strength of composites based on PP (MFR: 4.2 dg min^{-1}) and increasing amounts of talc. Injection molded samples tested according to ASTM D790. Data taken from Fujiyama [40]

cal upper crystallization temperature of the β-phase (i.e. 140 °C) and induce the irreversible growth of the α-modification. This effect is enhanced for low cooling rates and high amounts of talc. Combining active fillers with β-promoters is thus as meaningless as mixing β-PP with conventional α-nucleating agents [39].

A moderate content of β-modification was also observed in environmentally friendly composites, without promoting its formation by specific nucleating agents. Bamboo [153], woodflour [154], jute [155], and sisal fibers [156] were used as the fibers. We believe these components behave like inactive fillers, neither influencing the crystallization rate of PP nor promoting the growth of the α-phase. The β-form may arise from (by-)products of the reaction between specific additives of PP and the chemicals used to treat the surface of the fibers. Mechanical properties have only been reported for PP-sisal fiber composites; their stiffness and toughness increased linearly with their fiber content [156]. Although technical issues related to the intrinsic hydrophilic character of natural fibers remain to be solved, it would be worthwhile to study the selective promotion of the β-phase in such composite systems, especially because of their ecological impact.

A brand new area should conclusively be mentioned, that of the PP-nanocomposites made of single-walled carbon nanotubes [157] or silica nanoparticles [158] which were reported to facilitate the growth of the β-crystalline structure. However, no published data support their supposed outstanding performance.

5
Toughness in Rubber-Toughened β-Polypropylene

5.1
Rubber-Toughened PP: A Brief Overview

Rubber-toughened PP are heterophasic systems in which discrete elastomer particles are dispersed in a PP matrix. The rubbery phase enhances the impact resistance by initiating highly dissipative deformation mechanisms and by allowing the grades to develop up to the highest possible strains [159–162]. If the matrix is no longer able to accommodate the external load, as is the case at high speed and/or low temperature, these damage mechanisms are initiated by the particles (at least for a dilatational loading). Their role is threefold:

- promotion of plastic deformation of the matrix around the particles because of their role of stress concentrators;
- generation of cavitation phenomenon followed by a volume increase in the polymer, which allows an easier matrix flow;
- stabilization of the deformed polymer by a mechanism of craze deflection.

According to Partridge [163], toughening is efficient when, by comparison to the neat homopolymer tested under the same conditions, the impact resistance is multiplied by a factor of 10, without losing more than 25% of stiffness. The upper temperature limit for the use of rubber-modified blends is controlled by the matrix melt temperature, T_m, their lower limit by the glass transition temperature, T_g, of the particles. As soon as the viscoelastic response of the latter is too slow to accommodate an external loading, the polymer assumes a glassy state and breaks in a brittle way.

5.2
Influence of Rubber Content

Combining the benefits of both the rubbery phase and β-modification to boost in a super-proportional way the fracture resistance of PP grades is tempting, all the more that elastomers have been shown not to disturb the β-crystallization tendency of the matrix [10, 26, 38, 46, 119, 164–167]. Results from the pioneering work of Varga are summarized in Fig. 23 [165]. Up to 16 wt % of an EP-rubber (ethylene-propylene-rubber) were blended in a PP matrix with low flowability (MFR $\sim$ 1.5 dg min^{-1}). Flexural impact tests were performed on 3 mm-thick samples. As for non-nucleated systems, increasing (i) the rubber content and (ii) the temperature had a positive influence on the fracture resistance, because (i) each particle constitutes a potential site for initiation of plasticity and because (ii) matrix shearing is facilitated (towards) crazing at "higher" temperatures. In commercial systems the amount

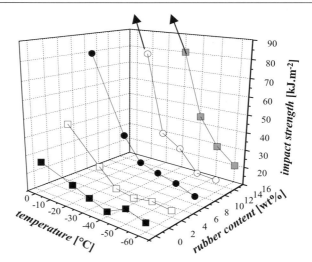

Fig. 23 Flexural impact strengths of β-modified rubber toughened PP plotted versus the amount of rubber content of the blends and the testing temperature. The *arrows* indicate samples that did not break. Data taken from Varga [165]

of rubber rarely exceeds 30 – 35 wt %, too high an addition of elastomer leading to an undesirable loss of rigidity. This competition stiffness/impact is illustrated in Fig. 24 for an injection molding propylene-ethylene block-copolymer (PP/EPR, MFR ~ 3.5 dg min^{-1}), which has been smoothly α-nucleated and strongly β-modified [124]. Although the data are thought to be close to the standard deviation of the measurements and should therefore be considered with caution, the E-moduli seem to be linearly but negatively

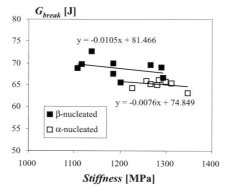

Fig. 24 Stiffness (ISO R527 with ASTM D638 type I samples) of α- and β-nucleated rubber toughened PP plotted versus their puncture impact resistance (ISO 6603-2) measured at 4 m s^{-1} and room temperature. Data adapted from Lambla [124]

correlated with the puncture resistances determined at $4\,\mathrm{m\,s^{-1}}$ on 2.8 mm-thick specimens. Moreover, the β-nucleated resin seems to be able to dissipate more energy at a given stiffness than its α-nucleated counterpart.

5.3
Influence of Rubber Molecular Weight

In the same way that the molecular weight of the matrix has a detrimental influence on the mechanical response of a grade, that of the rubbery phase, industrially described by the intrinsic viscosity, also plays an important role.

A model series consisting of six ethylene-propylene block copolymers (PP/EPR) has been investigated. While the flowability of their matrix (MFR:45 dg min^{-1}), the rubber content (about 23 wt %) and the amount of C_2 in the E/P rubber have been fixed, their intrinsic viscosity (IV) has been varied systematically from 1.7 to 6 dg l^{-1} (which corresponds roughly to M_w from 150 to 1000 kg mol^{-1}). A detailed description of the non-nucleated materials and their performance has been published elsewhere [168].

Fig. 25a provides a comparison of the Charpy notched impact strength (NIS) at room temperature of the non-nucleated and of the β-doped resins (fully β-modified with 0.1 wt % calcium pimelate) as a function of the IV of the rubber. Both series show the same trends: increasing the IV for the rubbery phase has a positive effect on the toughness. However, the delta between their respective fracture resistance at given IV, Δ(NIS), is not constant varying between 6% for an IV of 2 dg l^{-1} to 50% for an IV of 1.7 dg l^{-1}:

$$\Delta(NIS) = \frac{NIS(\beta\text{-nucleated}) - NIS(\text{non-nucleated})}{NIS(\text{non-nucleated})} \tag{8}$$

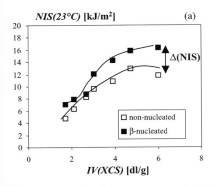
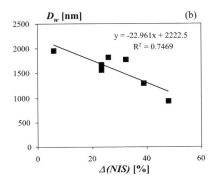

Fig. 25 a Notched impact strength (NIS) at room temperature of a non-nucleated and a β-nucleated PP/EPR model series plotted versus the IV of the rubber phase; **b** particle size, D_w, of the investigated series (from Grein et al. [168]) plotted versus the delta in NIS, at given IV, between the β-modified grade and its non-nucleated counterpart

These differences were correlated with the size of the rubber particles in the systems: the smaller the diameter of the dispersed phase (e.g. the lower the interparticular distance), the higher the benefits of a β-nucleation (Fig. 25b). For the grades with the smallest particle sizes, it might be attributed to an easier plastic deformation of the matrix once the damage mechanisms initiated (by particle cavitation) as a result of the smaller matrix ligaments between the rubber phase.

This knowledge constitutes a powerful tool for designing new classes of materials as patented by Bernreitner et al. [82].

5.4
Deformation Map of a Specific β-Nucleated PP/EPR

An accurate way to characterize the impact behavior of a material is to establish its deformation map. By analogy to the procedure used in Sect. 3.2.5 to described the fracture resistance of non-nucleated and β-nucleated PP of MFR12, a deformation map of their corresponding 15% modified blend (PP/EPR) is presented in Fig. 26. The temperature was varied in between – 30 and 60 °C and test speed in between 0.0001 and 10 m s^{-1} (but assessed preferentially in terms of crack tip loading rate, dK/dt). Tests were performed on 4 mm thick compact tension specimens. On the contrary to their neat corresponding matrix (whose fracture behavior has been discussed in Sect. 3.2.5), little difference could be recorded in their mode of failure. The β-nucleated PP/EPR were, admittedly, ductile up to slightly higher test speeds than their non-nucleated homologues, but far from the extent observed for the homopolymers. Their improved performance was, however, reflected by

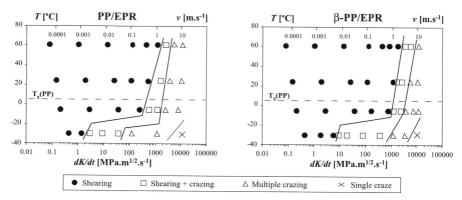

Fig. 26 Deformation maps of **a** a non-nucleated PP/EPR with 15% EPR and **b** its β-nucleated homologue for different temperatures and crack tip loading rates as deduced from the fracture surfaces of compact tension specimens. A rough indication of the test speed is provided by the upper scale

their initiation energies (G_{ini}, e.g. energy stored in the sample up to F_{max} in the F-d curve) which were for the whole tested range about 25–30% higher than those of the non-nucleated grades (Fig. 27). This feature was confirmed by their total energy to break (G_{tot}) highlighted in Fig. 28 for the measurements carried out at room temperature. The superiority of the β-nucleated resins is thus a consequence of the superior initiation of the damage mechanisms (G_{ini}) and of their better development in the volume ($G_{tot} - G_{ini}$).

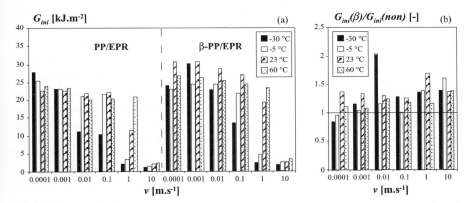

Fig. 27 **a** Initiation energy, G_{ini}, over the investigated temperature and test speed range of a non-nucleated PP/EPR with 15% EPR and its β-nucleated homologue; **b** ratios of G_{ini} as deduced from (**a**). Every point is the average of at least three independent measurements

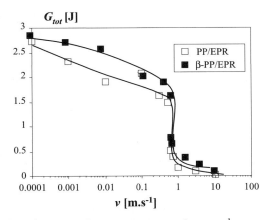

Fig. 28 Energy at break, G_{tot}, in between 0.0001 and 10 m s^{-1} at room temperature of a non-nucleated PP/EPR with 15% EPR and its β-nucleated homologue. Every point is the average of at least three independent measurements

5.5
Which Limiting Factor: Matrix or Rubber?

One of the fundamental clues in rubber-toughened systems is to know if the matrix or the rubbery phase is responsible for the unstable failure of a material. There are indeed two possibilities to explain a ductile-brittle transition:

- the damage mechanisms have been initiated by the rubbery phase, but could no longer grow in the matrix due to its insufficient ductility under the considered test speed and temperature;
- (almost) no cavitation could be initiated within the elastomer phase, rendering the growth of energy dissipative structures (e.g. crazes) in the matrix ligaments impossible.

We will limit our argument to the grades investigated in Sect. 5.4, but the way we propose to analyze the resin and conclude on their limiting factor can be applied to any material without restriction.

The observation of the fractured CT samples at room temperature provides valuable information about the damage mechanisms of the materials under investigation [169]. Two distinct whitened zones can be distinguished: one is intensively whitened (ZI), the other more diffusely (ZC). We infer ZI to be a zone where the damage forms can be initiated and propagated and ZC to be a zone where these deformation mechanisms are only initiated. The higher toughness of the β-nucleated grade towards its homologue is confirmed by the size of both its damaged zones over the whole studied range. At low test speeds, the deformation mechanisms are initiated and develop in the form of intense shearing and crazing in the matrix; at 0.001 m s^{-1} for example, ZI contributes to about 75% to the total whitened zone as obvious from Fig. 29. However, like G_{tot}, this zone becomes smaller with increasing test speed, to the benefit of ZC which grows continuously. Above the ductile-brittle transition (about 0.7 m s^{-1}), the damage mechanisms are no longer propagated in the continuous phase ($ZC \sim 0$), although they seem to be weakly initiated ($ZI \sim 1$ mm). The potential of the rubber phase therefore appears to be underexploited at high speed: the PP matrix seems to be the limiting factor.

This conclusion was only partly confirmed by scanning electron microscopy micrographs of RuO_4 stained surfaces taken at the crack tip of deformed specimens at 1 m s^{-1}, where the non-nucleated and β-nucleated materials showed, respectively, a semi-brittle and semi-ductile fracture behavior. While some limited rubber cavitation was visible for both resins, crazes—and consequently matrix shearing—could not develop to a large extent whether in the PP or in the β-PP matrix (although these structures were somewhat more pronounced in the latter case). Therefore, a question remains open: was the rubber cavitation sufficient to boost the development of dissipative mechanisms in these resins?

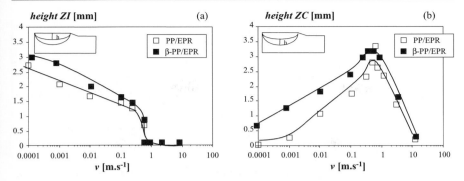

Fig. 29 Evolution at room temperature of the whitening of the fracture surfaces of compact tension specimens plotted versus the logarithm of the test speed for the non-nucleated and β-nucleated grades described in Sect. 5.4. **a** Height of the intense whitened zone (ZI); **b** height of the diffuse whitened zone (ZC) measured with a magnifying glass. The average standard deviation of these measurements is ±10%

To answer this question, a special testing procedure has been implemented. It allows the comparison of the resistance to fracture of both non-nucleated and β-modified grades, once the particle cavitation initiated. It consists of:

- initiating the damage mechanics by preloading a sample up to a given force. Concretely, the CT samples were deformed at 0.0001 m s^{-1} at 23 °C up to a displacement corresponding to F_{max}. The strong non-linearity observed in the F-d curve (see Fig. 17d) is associated with the development of a 3–5 mm whitened plastic zone in front of the crack tip;
- reloading the specimen in an instantaneous way ($\sim$ 500 mm min^{-1});
- evaluating the fracture behavior of the pre-deformed samples under pre-selected conditions. In this case, the measurements were carried out between test speeds of 0.0001 to 2 m s^{-1} at room temperature.

If the differences in toughness between both resins remain identical to the tests without preloading, the limiting factor is definitely the matrix; if not, the role of the rubbery phase cannot be neglected. The fracture energy of the preloaded specimens is given in Fig. 31. Considering that the transition from a stable crack growth to an unstable one is mirrored by the sharp decrease of G_{tot}, a shift of one decade of test speed is observed between the ductile-brittle transition of the β-nucleated grade and its non-nucleated counterpart. Keeping in mind that (almost) no difference was recorded for the non-preloaded samples, it can be inferred that the elastomer phase controlled the deformation process in these materials, although the matrix did not play a negligible role. A detailed study of these resins can be found in [77, 169].

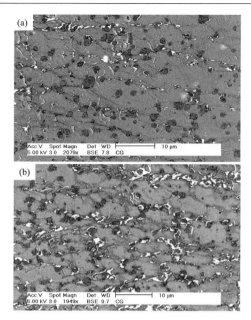

Fig. 30 SEM/BSE (Scanning Electron Microscopy/Back Scattering Mode) images of microdeformation of both **a** non-nucleated and **b** β-modified PP/EPR grades described in Sect. 5.4 tested at 1 m s^{-1} and room temperature

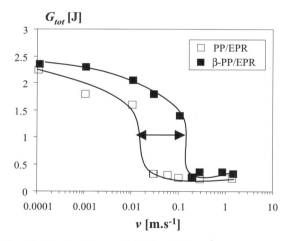

Fig. 31 Energy at break, G_{tot}, in between 0.0001 and 2 m s^{-1} at room temperature of a non-nucleated PP/EPR with 15% EPR and its β-nucleated homologue. The samples were preloaded samples at 0.0001 m s^{-1}. Every point is the average of at least three independent measurements

6
Specific Damage Mechanics in β-Nucleated PP

6.1
Some General Features

Relevant and complementary information about the damage process of polymers can be obtained among others by the analysis of the force-displacement curves, by the observation of the fracture surfaces (cf. Sects. 3.2.5 and 5.4) and, as will be shown in Sect. 6.2.2, by the determination of the amount of voids in a sample during and/or after deformation. However, a complete elucidation of the deformation mechanisms is only possible by their direct observation at the sub-micron level. Transmission electron microscopy is often used for this purpose. For convenience, the tests (which require experience and touch) are generally carried out at room temperature and at a low strain rate.

Under dilatational loading, the primary deformation modes are similar to those reported for other semi-crystalline polymers: separation of lamellae (followed by the formation of crazes), interlamellar shear and intralamellar slip [170–176]. Their relative intensity is dependent on intrinsic parameters, especially the molecular weight and stereoregularity of the investigated β-PP, and extrinsic factors like temperature, strain rate and temperature. In its early stages, the deformation is governed by the amorphous regions; in the later stages by the modification of the crystalline texture.

Interlamellar shear is an iso-volumetric process. It is present as the principal damage mode up to the beginning of cavitational deformation. Interlamella separation develops also in the early stages of the damage process, as soon as the material is no more able to accommodate a deformation at constant volume. It is typical for lamellae which are rather perpendicular to the loading direction, as confirmed by electron microscopy observations [173–175] and supported by theoretical considerations about the stress distribution in a deformed spherulite [176]. It is accompanied by an increase of the long period at constant lamella thickness [175] and leads to irreversible void formation (e.g. crazing). It competes, generally after necking, with the formation of chevrons which implies lamellae rotation (twisting) and break-up [173–176]. It has been reported to govern the damage process at high strain rates and/or when the material has a high degree of crystallinity and a high chain perfection. If this is not the case, interlamellar slip is believed to control the matrix flow. This phenomenon affects the zones where the lamellae are arranged along the loading direction and is thus believed to be the less active damage mechanisms in the first stages of the deformation process. Before necking, interlamellar slip is concentrated in the regions near the poles of the spherulites; it is accompanied by some lamellae thinning and in extreme cases by lamellae breaking. After necking, its operating range is extended to the lamellae, which rotate in the direction of the applied stress. These lamellae break either before or after rotation, constituting fragments separated by stretched amorphous macromolecules. Further deformation results in the creation of a highly fibrillated structure reflected macroscopically by strain-hardening. This process is accompanied by a phase transformation from (i) β to α or from (ii) β to the smectic form depending on the drawing temperature [128, 134, 136, 170–172, 177–179]. A high temperature is favorable to the formation of α-crystallites whereas a lower one (close to room temperature) rather promotes the development of a smectic phase [134, 170, 171, 177, 178]. None of the mechanisms proposed to explain the β/α phase transformation are commonly accepted. As a β/α solid-solid transition is not possible considering the spatial distribution of the chains

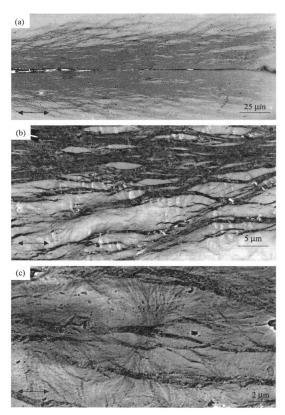

Fig. 32 SEM/BSE images of microdeformation of β-nucleated PP tested at room temperature and 0.001 m s^{-1} at the crack tip of a compact tension specimen: **a** overview of the crack tip damage zone, **b** and **c** details from the periphery of the damage zone. The *arrows* indicate the direction of the applied load

in PP [86], the formation of macrofibrils in the stress direction occurs most likely in a liquid phase by partial or complete melting of the β-lamellae and strain-induced recrystallization [55, 102, 170, 177, 178, 180, 181]. The α-nuclei necessary for such a growth has been suggested to be already present in the system: they are thought to arise from an epitaxial growth of the α-crystals on β-spherulites during a secondary crystallization process at temperatures below 100 °C [70, 122, 182].

The damage mechanisms of non-nucleated and β-modified PP differ in the localization and intensity of their primary deformation modes at given temperature and test speed. In ductile β-PP, the damage zone is rather homogeneous, extended and consists of a dense bundle of nearly parallel crazes which merges to form a continuous fibrillar structure adjacent to the crack faces (Fig. 32). These crazes develop either along the spherulites equators or in the corresponding polar regions, where the crystallites are roughly parallel to the loading direction [77, 173, 174, 176]. These non-isovolumetric structures are accompanied by shearing of the matrix; their respective amount will be quantified in section Sect. 6.2.1.

6.2
Some Experimental Findings

6.2.1
Enhanced Creation of Microvoids

The formation of microvoids (or micropores) in drawn β-PP is a well-known feature, used successfully to produce breathable films and membranes. Macroscopically their occurrence is correlated with an opacification (i.e. whitening) of the specimen. Their amount can, for example, be measured at given strain rate/draw ratio by liquid porosimetry (e.g. mercury intrusion), gas absorption or by the determination of the density of the deformed sample [129, 177, 178].

The content of micropores increases with increasing strain [177] up to a saturation level, lowering the drawing temperature [177] and increasing the crystallization temperature of the samples (in between 10 and 110 °C) [177, 178] as shown in Fig. 33. It can be further enhanced by drawing at constant width (e.g. forced superimposed volume increase) or under biaxial drawing [177]. The influence of the crystallization temperature has been explained by a competition between (i) the volume contraction induced by the β/α transformation (the density of the β-crystals being lower than that of the α-form, 0.921 against 0.936 g cm^{-3}); and (ii) the volume expansion induced by non-isovolumetric damage processes (e.g. crazing). For samples crystallized at low temperature (10 °C), on the one hand, the shrinkage consecutive to the rapid phase transformation caused by unstable β-crystals is believed to compensate a potential volume enhancement aris-

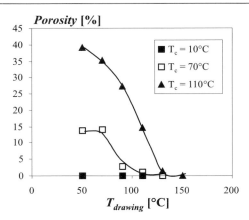

Fig. 33 Evolution of the amount of microvoids in 0.2 mm thin β-PP films obtained with different crystallization temperatures and deformed at temperatures ranging from 50 to 150 °C up to a draw ratio of 3.8. Data taken from Chu et al. [177]

ing from voids, leading macroscopically to no detectable change of volume in the deformed β-PP systems [177]. A similar phenomenon has been associated in an independent study with an enhanced microhardness for the deformed sample compared to the non-deformed one [183]. On the other hand, the β/α phase transformation has been proposed to be hindered by the inherent thermal and mechanical stability of β-lamellae crystallized at high temperature (110 °C), a feature which allows the development of numerous voids [177].

The decrease of the amount of microvoids with increasing temperature (Fig. 33) might be ascribed to a competition between shearing/crazing. Matrix shearing is indeed promoted at higher temperatures, the critical stress needed to induce plastic flow decreasing monotonically with increasing temperature. Following a similar train of thought, the content of micropores is expected to increase with increasing test speed. This fact was verified for non-nucleated and β-nucleated PP (MFR: 12 dg min^{-1}) tested between 0.0001 and 1 m/s at room temperature (i.e. with $0.001 < d\varepsilon/dt < 10$ s^{-1}) using a non-destructive on-line technique which separates the overall strain (ε) in a uniaxial tensile test into three independent and single contributions: shear, cavitational, elastic [77, 169]. As obvious from Fig. 34, the cavitational contribution, which accounts for dilatational processes, (i) was twice as pronounced for β-PP than for its counterpart (except at 10 s^{-1}); and (ii) increased markedly with the strain rate. Between 0.001 and 1 s^{-1}, the extent of cavitation at any given ε increases almost linearly with the logarithm of the strain rate, possibly indicating a thermally activated process. As these values have been determined for an elongation corresponding to the yield stress, it is indisputable that voiding precedes yield, as already suggested by Cheung [55]. Moreover, the onset of voiding/crazing of both grades has also been shown

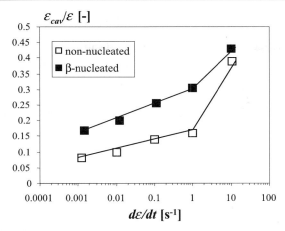

Fig. 34 Cavitational contribution, $\varepsilon_{cav}/\varepsilon$, of the overall deformation between 0.0001 and 1 m s^{-1} (i.e. at strain rates from 0.001 to 10 s^{-1}) for a β-nucleated PP tested at room temperature (MFR: 12 dg min^{-1}) and for its non-nucleated homologue

to decrease with increasing test speed from 7% at 0.001 s^{-1} to 1–2% at 0.1 s^{-1}, suggesting that above a certain strain-rate dependent threshold interlamellar shear can no longer accommodate the global deformation. In addition, as the elastic contribution has been shown to be substantially independent of the test speed, the cavitational and iso-volumetric processes (shearing) complement one another [77].

6.2.2
Dynamic Mechanical Analysis (DMA) Evidences

The intensity of the molecular relaxations of polymers have been shown to correlate positively with the toughness since a higher chain mobility (reflected by a higher strength of the relaxations) (i) reduces the elastic energy stored at a given strain and stress and delays thus the occurrence of crazing; (ii) lengthens the phase of growth of the crazes contributing to hold up the development of an unstable crack; and (iii) allows the dissipation of more energy during the propagation of the damage mechanisms [184–187]. Polypropylene exhibits three molecular relaxations independently of its crystalline modification [188, 189]:

- a γ-relaxation at – 70 °C, attributed to the rotations of side methyl groups in the glassy stage;
- a β-relaxation around 0 °C, accounting for the glass transition of the amorphous PP macromolecules;
- an α_c-relaxation which begins at about 50 °C and is generally considered to describe the diffusion of crystallographic defects in the lamellae.

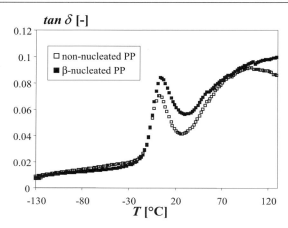

Fig. 35 DMA trace of non-nucleated and β-nucleated PP measured on 1 mm thick compression molded plates. Tests have been carried out in a torsion mode at 1 Hz with a heating rate of 1 K min^{-1}

For the DMA trace in Fig. 35—typical for almost all reported literature data—it is obvious that both α_c- and β-relaxations are more pronounced for the β-nucleated material than for its non-nucleated counterpart suggesting both its amorphous and crystalline phases to be more mobile. Varga showed the progressive increase of the maximum of the loss factor, tan δ, as K_β was progressively increased from 0 to 75% [137]. This improved damping obviously favors fracture resistance [48, 72, 77, 190]. Detailed correlations between the intensity of the relaxations and the impact strength in β-nucleated PP would be beyond the scope of this review; they are published elsewhere [111, 191].

A careful observation of the DMA traces leads to further findings. Labour [140] has reported a gradual shift of the α_c-peak to lower temperature by increasing the β-content of the PP (between 0 and 31%) suggesting an easier activation of the chain mobility in the β-crystals than in the α-ones. Moreover, it has been pointed out that there is a slight shift to a higher temperature for the glass transition of β-PP compared to non-nucleated PP (as also visible in Fig. 35). This phenomenon has been attributed to a slight immobilization of the amorphous phase in β-PP, which has been proposed to be a result of its numerous tie-molecules.

6.2.3
Macroscopic Tensile Indicators

Besides the microscopic damage forms and the molecular relaxations, macroscopic indicators also provide interesting hints of the deformation processes of PP. The yield stress (σ_y) is significantly lower in the β-nucleated resins than in their unmodified homologues (Fig. 36), reflecting their higher ability to initiate plastic flow at lower stresses [48, 55, 64, 76, 77, 84, 109, 124, 137, 140, 172, 178, 179, 190]. It can be described by a two stage Eyring equation (Eq. 9) [77, 141, 142]:

$$\frac{\sigma_y}{T} = \frac{R}{v_1}\left[\frac{\Delta H_1}{RT} + \ln\frac{2d\varepsilon/dt}{d\varepsilon/dt_{01}}\right] + \frac{R}{v_2}\sinh^{-1}\left[\frac{2d\varepsilon/dt}{d\varepsilon/dt_{02}}\exp\frac{\Delta H_2}{RT}\right] \qquad (9)$$

where σ_y is the yield stress, T is the test temperature in Kelvin, ΔH_i is the activation energy of the plastic flow for the mono-activated process i, v_i is the activation volume of the element motion unit for the process i, $d\varepsilon/dt$ is the strain rate and $d\varepsilon/dt_{0i}$ is a pre-exponential factor for the process i.

Considering the low test speed yield stress required to be a mono-activated process (the second process being considered to be predominantly active at a higher strain rate), the activation volume of both crystalline modifications of PP estimated from the slope of the σ_y-$d\varepsilon/dt$ curves in the different regions is not significantly different, but slightly higher for the β-form in accordance with literature data [48, 55, 179]. The values are around 4 nm^3 for strain rates up to 1 s^{-1} and around 3 nm^3 for strain rates between 1 s^{-1} and 100 s^{-1}. Although the physical meaning of v is unclear, one may suspect the yield stress at the higher test speed to be controlled to a larger extent by the crystal motions activated at – 70 °C. Moreover, the plastic flow seems to involve cooperative motions of PP segments rather than cooperative motions of chains.

The Young's modulus, E, of β-PP and its blends is for ideally nucleated grades about 15% lower than their corresponding non-nucleated grades (Table 5) as reported in the literature [32, 38, 48, 76, 84, 109, 124, 137, 169, 190]. This feature may originate from the same factors that control the yield stress: (i) a lower cohesive force of the β-phase, as a consequence of its lower pack-

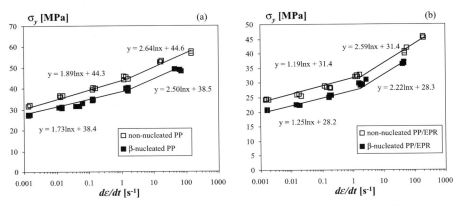

Fig. 36 Typical evolution of the yield stress, σ_y, of **a** non-nucleated and β-modified PP and their **b** 15 wt % toughened PP/EPR blends as a function of the strain rate, $d\varepsilon/dt$. Tensile tests performed at room temperature on a base resin exhibiting an MFR of 12 dg min^{-1}

Table 5 Young modulus, E, for non-nucleated and β-modified PP and their 15 wt % toughened PP/EPR blends as a function of the strain rate, $d\varepsilon/dt$. Tensile tests performed at room temperature on a base resin exhibiting an MFR of 12 dg min^{-1}

	E at 23 °C [MPa]
Non-nucleated PP	$E = 102.4 \ln(d\varepsilon/dt) + 2264$
β-Nucleated PP	$E = 78.4 \ln(d\varepsilon/dt) + 1774$
Non-nucleated PP/EPR	$E = 82.6 \ln(d\varepsilon/dt) + 1875$
β-Nucleated PP/EPR	$E = 70.8 \ln(d\varepsilon/dt) + 1591$

ing density; and (ii) a favorable lamella arrangement (no cross-hatching). An additional fact which may explain the higher toughness of the β-phase is its potential to initiate non-linear processes (either delayed elasticity or plasticity) at lower stresses than its non-nucleated homologue [109, 169]. For example the elasticity limit (E_{limit}, in MPa) of CaCO$_3$ filled PP, determined in plane strain compression at 0.001 s^{-1}, has been shown to decrease roughly monotonically with the β-content (K_β varying from 6 to 41%, expressed in % in Eq. 10):

$$E_{\text{limit}} = -0.243 K_\beta + 58 \tag{10}$$

As this phenomenon is more pronounced at $T > T_g$ than at temperatures below T_g, the crystalline phase rather than the mobility of the amorphous phase has been proposed to control the onset of non-linearity [140].

A further difference between both α and β-modifications is the extent of strain-hardening they exhibit. Convergent sets of data highlight the more prominent strain-hardening of the β-phase in the post-yield range compared to its non-nucleated counterpart [109, 140, 172, 179]. This fact results most probably from (i) the easier plastic deformation of the β-crystals and of (ii) the transformation into the fibrillar structure at earlier stages than for the α-phase.

Finally, the delocalization of the damage mechanisms is much more important for β-nucleated resins independent of their nature (homopolymer or copolymer) than for their non-nucleated homologues [48, 77, 109, 172, 176, 177, 179]. This can be rationalized in terms of the difference between the plateau stress and the yield stress, Δ_{py}, in stress-strain engineering curves: the higher this difference is, the higher is the localization of the necking [169]. As obvious from Fig. 37, the β-nucleated plastic zone is much more extended than that of the non-nucleated corresponding grades, the consequence of a less pronounced necking and of a better distribution of the input energy in the tested sample. Moreover, Δ_{py} increases linearly with the logarithm of the strain rate, but decreases in a less monotonic way with the temperature, indi-

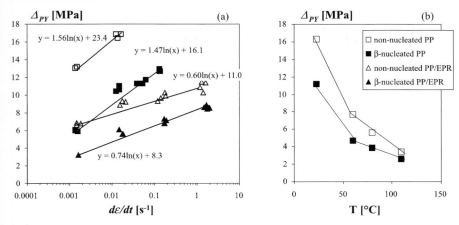

Fig. 37 a Difference between the plateau stress and the yield stress in stress-strain engineering curves, Δ_{py}, of β-nucleated homopolymers and PP/EPR blends (15 wt % EPR) and their corresponding non-nucleated homologues plotted as a function of the strain rate. Tests performed at room temperature on injection molded dog-bone specimens on PP (homopolymer and matrix of the copolymer) having an MFR of 12 dg min^{-1}. **b** Δ_{py} of non-nucleated and β-nucleated homopolymers (MFR: 1.7 dg min^{-1}) plotted as a function of the temperature. Tests performed at 100 mm min^{-1} on injection molded specimens. Data taken from Fujiyama [172]. The legend is the same for both parts of the figure

cating a less pronounced necking (e.g. a better delocalization of the damage mechanisms) under smooth test conditions.

6.3
What Explains the Toughness of β-Modified Polypropylene?

6.3.1
Lamella Architecture

The facts presented in Sect. 6.2 highlight the complex relationships existing between the crystalline and amorphous components of the β-modification. On the one hand, evidence for the determining role of the crystalline phase has been seen. This has been used to explain the evolution of the macroscopic tensile indicators of β-PP as a function of the test speed and temperature. On the other hand, the increased mobility of its amorphous phase, reflected by the DMA trace, cannot be neglected. Both factors can be reconciled considering the arrangement of the β-lamellae. Compared to the α-phase, β-PP exhibits a sheaf-like structure of radial lamellae growing in bundles from a central nucleus without any epitaxial growth of tangential lamellae. As a consequence, the development of extended plastic damage forms will not be disturbed by the physical network created by cross-hatched crystallites. Concretely, the deformation of the amorphous chains (and therefore the lamella separation process) will be facilitated in the early stages of a dilatational deformation for the β-modification since no "interlocking" structure blocks their mobility, which allows an efficient stress transfer [72, 75, 77, 78, 137, 140, 176]. An experimental proof for this feature has been given in Sect. 6.2.1: the amount of microvoids measured in β-PP was definitely more important than that determined for the corresponding non-nucleated PP. The enhanced strength of the β-relaxation for β-modified PP grades, observed in DMA, also suggests this scenario, especially because this technique reflects viscoelastic behaviors at small strains. At large strains, plastic slip processes constitute the main damage mechanism at lower strain rates. They are accompanied by crystalline dislocations which move on preferred glide planes, those of critical resolved shear stress. In the α-phase, a certain correlation is needed between the plastic events occurring in neighboring lamellae to allow the creation and propagation of these dislocations as remarked by G'Sell [176]. Due to their cross-hatched structure, the movement of glides between radial and tangential lamellae is believed to be difficult. On the contrary, crystals can deform more independently in time and space in the β-modification. Propagation of dislocations and crystal slip are thus easier, and all the more facilitated that the β-modification exhibits a hexagonal structure with three equivalent glide planes. As a result, the crystals arranged statistically in a favorable orientation (with regard to the principal shear stress) are expected to be more numerous and the deformation more uniform.

6.3.2
Additional Factors

Besides the fundamental differences in their lamellae architecture, additional factors may further explain the toughness behavior of both non-nucleated and β-modified PP.

One of them is the spherulite size [50, 77]. This parameter is in first approximation induced by the mode of crystallization of both modifications. Whereas non-nucleated PP crystallizes in a homogeneous way, the nucleation process in β-PP is heterogeneous, the nucleating agents acting as germination sites. In this latter case, the density of nuclei is 10 to 100 times higher than with a non-nucleated homopolymer system, leading to 10 to 100 times lower spherulitic sizes (1–10 μm for β-PP towards 30–100 mm for neat PP). As smaller spherulites are known to promote toughness, especially at low speed and/or high temperature, owing to the reduced concentration of structural defects and impurities at their boundary, part of the difference in fracture resistance between both crystalline structures is to ascribe to the size of their respective superstructures. However, its effect is limited; although in studies which erased in a defined way the differences in spherulitic sizes, strong differences could be seen between non-nucleated and β-modified systems [75, 76, 140, 146].

The lower packing density of the β-crystalline form is also suspected to contribute positively to the toughness [75, 140, 172]. As already outlined in Sect. 6.2.3, the induced reduced chain interactions may promote an easier plastic deformation of the lamellae (particularly a favored slip of the lamellae chains), as a consequence of the lowered energy barrier of the most probable conformational defects (e.g. 120 chain twist accompanied by a c/3 shift in the stem direction) as suggested Labour et al. [140].

The suspicious (but experimentally well reproducible) maximum found in the fracture resistance of NU-100 β-nucleated PP for concentrations of about 300–500 ppm (see Sect. 3.2.1) has oriented the recent research in a rather unexplored direction: the optimization of the β-structure itself. Raab [84] has suggested that a bundle-like morphology without a real spherulitic structure exhibits a higher plasticity and toughness than well-developed spherulites. In cooperation with Grellmann, he has also associated the best performing β-PP—in a series where the concentration of the nucleating agent was varied systematically—with a maximum in thickness of the amorphous interlayers (expressed in terms of the maximum of the long period, LP) and with a minimum in the lateral size, L_{300}, of the crystallites. Results are addressed in Fig. 38 [64]. Also the evolution of the dynamic fracture toughness, J_{Id}, rather than that of the static elongation at break, ε_{break}, correlates with the evolution of LP and L_{300}. An alternative or completive explanation for the toughness optimum observed with NU-100 has been proposed by Obadal [68]. It is based on the relative amounts of the β_1 and β_2 melting peaks (recorded by DSC) over the thickness of the sample (measured each 100 µm!). Whereas the

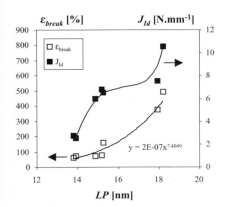
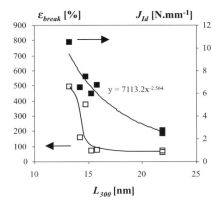

Fig. 38 Evolution of J_{Id}, the J-Integral accounting for the fracture toughness, and of the elongation at break, ε_{break}, as a function of the long period, LP, (*left*) and the lateral size of β-crystals, L_{300}, (*right*). Tests have been performed at room temperature on injection molded samples at 1.5 m s^{-1} to determine J_{Id} and at 50 mm min^{-1} to measure ε_{break}. Mechanical and structural responses have been varied by varying the concentration of NU-100, the nucleating agent from 0 to 1300 ppm

region close to the skin is enriched in β_2-phase that nearer to the core contains a larger amount of β_1-modification. Calculating the strength of β_2/β_1 from the skin to the core, it is possible to determine a critical thickness for which $\beta_2 > \beta_1$. The maximum in Charpy notched impact strength observed for NU100 concentration ranging from 0 to 2000 ppm has been associated with a maximum in this critical thickness. Interestingly, the authors stated that β_2 is connected with the relaxations of elongated chains (e.g. not with a transformation from an unstable structure, β_1, to a more stable one, β_2, during heating).

6.3.3
The Controversial β-α-Phase Transformation

One cannot conclude this section devoted to the mechanisms or factors of toughness improvement without evoking the β/α transformation. Basically, the occurrence of a β/α transformation during loading has been largely documented and is incontestable [55, 72, 78, 128, 170–173, 177–179, 192]. What remains unclear is if this transformation is the driving force of the outstanding mechanical performance of β-nucleated grades or only its consequence.

The followers of a phase transformation toughening (PTT) state that the change from a less densely packed crystalline structure (β) to a more packed one (α) (i) promotes microvoiding in the earliest stages of the deformation and (ii) facilitates plasticity due to its exothermic character [72, 78, 193].

The more reserved opposing party argues that the temperature increase is certainly due to a self-heating phenomenon resulting from the adiabatic transformation of the plastic work into thermal energy and attributes the enhanced amount of microcrazes observed in β-PP to its lamellae architecture [75]. Some further arguments might support their opinion. A β/α transformation has not been observed systematically after dilatational testing [50, 76, 77]. Especially in confined structures (e.g. crack tips of notched specimens), no experimental evidence could be provided up to now for the occurrence of such a transformation either because it is too localized to be detected or because it does not exist. It has also been reported by two independent groups that no detectable conversion from β to α occurs before a draw ratio of 1.2 (at 23 °C) [67] and 1.6 (at 100 °C) [55]; after these onsets, the β-content decreases monotonically with increasing strain. Moreover, the deformation of β-PP at low temperature has been shown to lead to a smectic (or α-form with low crystallinity phase [172]), rather than to a high crystalline α-modification [128, 133, 134, 177, 178]. In addition, it should be remarked that the transformation between two such different systems is not possible without rewinding of the chains and should therefore take place more likely in a liquid phase. Even if such a transformation would concern individual lamellae, one might doubt that a local heat increase at low strains is important enough to allow it.

However, whatever the real importance of the β-α transformation for the improved toughness of β-PP, it is greatly to its credit that it has opened up new ways of thinking in the β-PP world and contributed to impassioned scientific debates.

7
Some Applications of β-Nucleated Polypropylene

7.1
Fibers

In spite of the increasing use of PP in the textile industry, the selective modification of its crystalline structure by β-nucleating agents remains confidential. Besides its inherent (and desired) coloration effect when working with γ-quinacridone [194], it improves the poor moisture absorption of PP fibers [128, 129, 132–134]. This feature is obtained by introducing, in a targeted way, micropores in the fibers during the drawing process which follows melt spinning. It is accompanied by a progressive disappearance of the β-form to the benefit of the α-phase. Working with a fully nucleated β-PP and choosing the right drawing ratio and temperature are the keys to success for such an application.

The β-content of the fibers can be optimized by adjusting the spinning and subsequent cooling parameters. To allow the growth of a well-developed crystalline β-phase, a low cooling rate – and thus a higher temperature of the cooling medium and/or a cooling medium with low thermal conductivity (rather air than water)—is preferred [129]. The choice of the right spinning speed (e.g. take-up speed) and spinning temperature is less straightforward: evolutions of the β-content in opposite directions have been reported by varying them [128, 129, 132–134]. These parameters are believed to be dependent on each production unit (especially on the distance between the spinning nozzle of the extruder and the take-up roll) and should therefore be evaluated case by case. Moreover, it has been reported that thinner fibers are less prone to develop a β-form [128, 129], information which is of limited importance in practice.

The optimum draw ratio is close to the micropore saturation level (if any), since further drawing would lead to an undesirable embrittlement of the fiber without substantial gains in the amount of microvoids. By analogy with the results obtained with stretched films, a maximum of micropores is expected for drawing temperatures around 80–110 °C [128, 129]. However, no incontestable experimental evidence has been provided up to now to support this assertion.

Besides these high moisture absorption fibers, a new application field has emerged [195]. An all PP solution consisting of a β-nucleated matrix and fibers of α-crystalline structure has been patented. It takes profit of the lower melting temperature of β-PP towards α-PP, which—because it allows an easier incorporation of the fibers into the continuous phase—facilitates the production of such a composite. In this application, β-PP constitutes an interesting alternative to metallocene PP or C2-rich random copolymers which could have also been used as a matrix. Such a α-fiber/β-matrix composite has been claimed to be competitive with glass-fiber reinforced PP.

7.2
Glass-Fiber Reinforced Polypropylene

The most exotic application associated with β-nucleating agents is certainly the "β-doped glass-fiber reinforced PP" [196]. It deserves, however, to be mentioned in this review since it constitutes an original approach (i) to take profit from both α- and β-modifications of PP within the same material and (ii) to improve simultaneously both stiffness and impact resistance. It consists of a fiber/PP composite with a β-PP interface between the dispersed phase and non-nucleated PP phase. The β-transcrystallinity has been introduced by coating the glass fibers with a highly active β-promoter. For a 40 wt % filled composite, an increase of about 13% in both stiffness and toughness could be measured between the non-nucleated grade (resp. 4450 MPa for the E-modulus and 88 kJ m^{-2} for the Izod Impact Strength) and the β-doped composite (resp. 5082 MPa and 102 kJ m^{-2}) [196].

In "single fiber composite" tests it could be shown that for β-transcrystallinity, only interlamellar failure is observed without any evidence of bulk damage (in contrast to what was observed with α-transcrystallinity) [149–151]. It was attributed to the inherent tougher fracture behavior of the β-phase (towards its α-homologue) and to the position of the β-lamellae on the fibers: they lay "flat-on" e.g. with the lamellar surface parallel to the length of the fiber. For a composite containing the α-form as the continuous phase and the β-phase as a transcrystalline layer between fiber and matrix, scanning electron microscopy has revealed that cracks, initiated in the α-regions, propagated until the β-boundary. There, the cracks seems to be stopped or at least somewhat delayed since the damage forms observed in these β-rich regions seems to be more uniform and well distributed than those observed in the α-matrix.

7.3
Thermoforming

Thermoforming is a process in which extruded sheets are converted to articles by heating above their softening point, forming the softened sheet in a mold and allowing the formed sheet to cool and harden. The typical MFR range for such products is from 1 to 5 dg min^{-1}, typical applications are

Table 6 Comparison of some relevant properties in thermoforming between β-nucleated PP and other classes of polypropylene. 0 means identical to PP-homopolymer without nucleating agent; + means good, ++ very good, - bad, – very bad. The data provided for PP block and random copolymers deal with classical resins. Data taken from Wolfschwenger et al. [49]

	PP-homopolymer			PP block copolymer Non-nucleated	PP random copoylmer Non-nucleated
	Non-nucleated	α-Nucleated	β-Nucleated		
Stiffness	0	+	+	–	–
Cycle time	0	–	++	+	+
Heat resistance	0	+	+	–	–
Transparency	0	+	+	–	+
Toughness	0	0	0	+	+

food containers. The benefits of β-modified PP (and its blends) compared to non- or α-nucleated PP are both economical and technical [49, 117, 197]. Because of its low melting temperature (T_m: 145–150 °C), faster production rates are achieved. As the thermoformed article cools, the molten part of the PP recrystallizes in an α-modification so that there is almost no more β-phase in the thermoformed articles. As a consequence, no drawbacks are expected with β-nucleated grades regarding their stiffness, heat resistance and transparency. A comprehensive summary of their performance in comparison with that of different classes of PP is given in Table 6. A less common end use property, microwave-qualification, has also been reported to be improved starting from β-extrusion sheets [197]. Another important advent for β-PP in this segment is the (quasi)-absence of built-in stresses in the thermoformed parts as well with Ziegler–Natta as with metallocene-catalyst-based resins [117].

7.4
Microporous Films

The segment where β-nucleated PP has found its most fertile ground is that of microvoided films. The superior ability of the β-modification to create micropores during a non iso-volumetric damage combined with its excellent toughness has been exploited. Potential applications include filters, breathable membranes, absorbing articles, printable films, battery separators, dielectric capacitors, clothing materials for footwear such as leather substitutes or rainwear [19, 59, 60, 130, 198, 200–209]. Since structures that enable vapors to flow through them (i.e. are breathable) while at the same time inhibiting or stopping the flow of liquids through them are desired, the microvoids have to be interconnected. A classical MFR range of the β-PP (and eventually its blends) used for such applications is 2 to 8 dg min^{-1}; they may contain chalk to further improve their breathability.

Four different ways to achieve the targeted porosity have been mentioned. The common step involved consists of forming a film/sheet with a defined (and generally high) amount of β-modification, preferably with selective β-nucleating agents. Their critical parameter is the cooling rate of the film knowing that: (i) under isothermal crystallization little to no β-crystallites are formed below about 80 °C and above about 140 °C; (ii) if the film is cooled too quickly (or quenched) no β-phase might be formed; (iii) if the film is cooled too slowly a mixture of α- and rather big β-spherulites develops. As a result, the cooling rate is used as a design tool to control the amount and size of the β-spherulites in the film. A decrease in the cooling rate (and thus an increase in the β-spherulite size) is expected in extruded films when at least one of the following factors is increased: polymer melt, extrusion rate, die gap, cooling air temperature and chill roll temperature. Specific strategies follow this stage. The first one consists of melting the β-spherulites by heating the film in between $T_\mathrm{m}(\beta\text{-modification})$ and $T_\mathrm{m}(\alpha\text{-form})$ and stretch it [208, 209]. It has been extensively used to roughen BOPP films (Biaxial Oriented PP). The second strategy is the most common one. A film is heated between 30 and 135 °C and stretched [30, 59, 60, 200, 204–210]. Compared to non-nucleated PP blended with an inorganic resin (which may or may not be extracted before stretching), the main advantages of this technique are: (i) a homogeneous opacity; (ii) the easiness with which thin films are processed; and (iii) a high elongation at break. The third and forth way to get breathable and waterproof articles, patented by Jacoby et al [206, 207], are correlated. Their use depends on the composition of the system, essentially on the amounts of rubbery phase and low molecular weight PP. A selective extraction of the β-spherulites with a non-polar organic solvent, preferably toluene at about 90 °C, is in both cases performed. This might be followed by a stretching at about 110–135 °C, if the number, form and size of the voids formed during the removal of the β-phase is not optimal. Films formed in this way exhibit good strength and ductility.

7.5
Piping Systems

Homopolymers and copolymers with a strong β-modification offer many advantages for pressurized pipe applications, especially for chemical dispensing and industrial sewerage systems [61, 73, 80, 211–215]. Their high impact strength, long lifetime, easy handling and jointing, excellent resistance to chemicals and wide operating window outperform alternative materials. They are used as a unique material or in combination in multilayer systems.

In terms of toughness, one should distinguish (i) the long-term performance assessed by the resistance to slow crack growth (SCG); and (ii) the rapid crack propagation (RCP) level (also known as fast brittle fracture), which describes the ability of a grade to trigger unstable crack growth under static or, more frequently, dynamic loading.

Rapid crack propagation (RCP) is characterized by (i) high crack speeds (typically 100 ms^{-1} and more); (ii) smooth fracture surfaces with extremely limited plastic deformation; and (iii) a wavy propagation of the crack along the extrusion direction. RCP is the consequence of the local exceeding of a critical stress, energy or pressure. It may be caused either by internal defects (poor welding, fatigue crack of critical size) or external impact damage during service. Rapid crack propagation can be described with adapted LEFM parameters, the dynamic fracture resistance, G_D, and the dynamic fracture toughness, K_D. Their estimation implies the determination of the critical pressure, p_C, at which the crack propagates instantaneously in the system; the higher this pressure, the higher both K_D and G_D are [216]. Compared to parameters generated from perforation tests on finished parts or standard dynamic fracture resistance measurements, K_D and G_D are geometry independent and can thus be introduced in calculation codes. The superiority of β-nucleated PP and its blends is obvious: K_D and p_C are three times higher with such grades than with a random copolymer of comparable MFR (p_C = 20 bars/K_D = 4.1 MPa m$^{1/2}$ towards p_C = 8 bars/K_D = 1.5 MPa m$^{1/2}$ [216]). Another important safety parameter is the temperature at which the transition from an unstable to a stable (T_{db}) mode of failure occurs under a given pressure. For a β-nucleated PP, T_{db} is about 23 °C, for a random PP about 40 °C, both grades being devoted to pipe applications. Consequently, a damage map as a function of the temperature and pressure can be drawn for each system under investigation (Fig. 39); the higher p_C and the lower T_{db}, the more resistant the grades to brittle fast fracture. Moreover, due to the homogeneous and fine structure of β-spherulites the frozen-in-stresses in pipes are

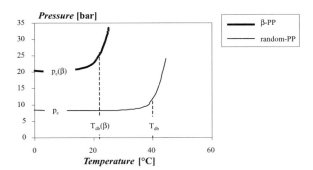

Fig. 39 Deformation map as a function of the temperature and the pressure of a random copolymer and a β-nucleated PP. Data adapted from Greenshields [216]

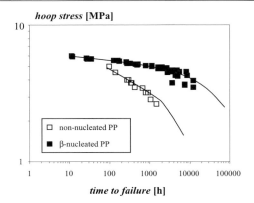

Fig. 40 Hoop stress plotted as a function of the testing time for both non-nucleated and β-nucleated PP (resp. BE50 and BE60, Borealis grades). Tests performed at 95 °C

significantly lower, an additional factor which aids their high stability towards external stresses.

Besides rapid crack propagation, slow crack growth (SCG) is the second important safety and design criteria in pipes. It is traditionally assessed by the long-term hydrostatic strength (e.g. hoop stress). The higher the hoop stress and the flatter its evolution as a function of time, the more resistant is the material to slow crack propagation. Fig. 40 highlights the excellence of a β-nucleated PP towards its non-nucleated homologue, both grades having an MFR of 0.3 dg min^{-1} usual for pipe applications. Micromechanically, SCG as a unique factor of this review is driven by a competition between crazing by disentanglement (at lowest test speeds and highest temperatures) and shearing (under somewhat higher test speeds and lower temperatures), all other parameters being controlled by a competition between shearing and crazing by scission (which has been designed as "crazing" in this article). This is, among others, the reason why (i) slow crack growth can only be simulated by time-consuming long-term tests; and why (ii) care should be taken by extrapolating data obtained in one temperature range into another since the dominant damage mechanisms in both ranges are not necessary the same.

8
Conclusion

This review has highlighted several aspects of the toughness of β-nucleated PP. Although its long-term behavior because of its relevance in pipe applications has been outlined, the focus has been on dynamic fracture properties, a topic which has been largely documented in the literature. Ductile-brittle fracture maps as the basis of the understanding of the mechanical performance of β-modified grades have been established. The influence of intrinsic parameters like molecular weight, polydispersity and matrix randomization has been discussed, that of extrinsic factors like stress-state, processing conditions, test speed and temperature illustrated. All of them have been shown to play a non-negligible role in the macroscopic mechanical response of the materials. Except for long-term properties, two damage processes have been shown to govern the fracture resistance of the β-PP: crazing by scission at low temperature and high speed and matrix shearing at lower strain rate and higher temperature.

Under defined conditions, the toughness is also driven by the content and spatial distribution of the β-nucleating agent. The increase in fracture resistance is more pronounced in PP homopolymers than in random or rubber-modified copolymers. In the case of sequential copolymers, the molecular architecture inhibits a maximization of the amount of the β-phase; in heterophasic systems, the rubber phase mainly controls the fracture behavior. The performance of β-nucleated grades has been explained in terms of smaller spherulitic size, lower packing density and favorable lamellar arrangement of the β-modification (towards the cross-hatched structure of the non-nucleated resin) which induce a higher mobility of both crystalline and amorphous phases.

The damage process is accompanied by numerous microvoids, the development of which has been utilized for waterproof and/or breathable films. Other interesting segments are fibers, glass-fiber reinforced PP, thermoformable grades and piping systems.

Acknowledgements The writing of this review would not have been possible without the support of Dr. Kurt Hammerschmid (Borealis). His scientific assistance and day-to-day presence have been precious allies. Special thanks also to my colleagues DI. Klaus Bernreitner, Dr. Markus Gahleitner, Ing. Johannes Wolfschwenger, Dr. Wolfgang Neißl (all Borealis) for the stimulating discussions we have had concerning the β-topic. I would also like to point out the decisive role of my PhD stay at the Swiss Federal Institute of Technology (EPFL, Lausanne) as the starting point of my interest in β-nucleated systems. I therefore specially acknowledge Prof. Hans-Henning Kausch, Dr. Philippe Béguelin, Dr. Christopher J.G. Plummer and Dr. Rudolf Gensler.

References

1. Crissman (1969) J Polym Sci 7(A2):389
2. Fujiwara Y (1975) Colloid Polym Sci 253:273
3. Lovinger AJ, Chua JO, Gryte CC (1977) J Polym Sci Polym Phys Ed 15:641
4. Leugering HJ, Kirsch G (1973) Angew Makromol Chem 33:17
5. Devaux E, Chabert B (1991) Polym Commun 32:464
6. Devaux E, Gerard JF, Bourgin P, Chabert B (1993) Compos Sci Technol 48:199
7. Varga J, Karger-Kocsis J (1993) Compos Sci Technol 48:191

8. Varga J, Karger-Kocsis J (1993) Polym Bull (Berlin) 30:105
9. Varga J, Karger-Kocsis J (1995) Polymer 36:4877
10. Varga J, Karger-Kocsis J (1996) J Polym Sci Part B: Polym Phys Ed 34:657
11. Jay F, Haudin JM, Monasse B (1999) J Mater Sci 34:2089
12. Ellis G, Gomez MA, Marco C (2004) J Macromol Sci Phys 43:191
13. Leugering HJ (1967) Makromol Chem 109:204
14. Kathan W (1986) DE Patent 3 443 599
15. Garbarczyk J, Paukszta D (1985) Colloid Polym Sci 236:985
16. Marcincin A, Ujhelyiova A, Marcincin K, Alexy P (1996) J Therm Anal 46:581
17. Ikeda N, Yoshimura M, Mizoguchi K, Kitagawa H, Kawashima Y, Sadamitsu K, Kawahara Y (1993) EP Patent 557 721
18. Huang M, Li X, Fang B (1995) J Appl Polym Sci 56:1323
19. Wolfschwenger J, Bernreitner K (1995) EP Patent 682 066
20. Kawai T, Iijima R, Yamamoto Y, Kimura T (2002) Polymer 43:7301
21. Fujiyama M (1996) Intern Polym Process 11:271
22. Ye C, Liu J, Mo Z, Tang G, Jing X (1996) J Appl Polym Sci 60:1877
23. Li JX, Cheung WL (1997) J Vinyl Additive Technol 3:151
24. Kobayashi T, Killough L (1997) AddCon Asia '97, Int Plastics Additives and Modifiers Conf, Singapore
25. Berson AL, Claverie F, Drujon X, Lotz B, Wittmann JC, Thierry A (1998) EP Patent 887 375
26. Varga J, Mudra I, Ehrenstein GW (1999) J Appl Polym Sci 74:2357
27. Marco C, Gomez MA, Ellis G, Arribas JM (2002) J Appl Polym Sci 86:531
28. Li X, Hu K, Ji M, Huang Y, Zhou G (2002) J Appl Polym Sci 86:633
29. Feng J, Chen M, Huang Z, Guo Y, Hu H (2002) J Appl Polym Sci 85:1742
30. Busch D, Kochem K, Schmitz B, Tews W (2003) WO Application 2 003 091 316
31. Maeder D, Hoffmann K, Schmidt HW (2003) WO Application 2 003 102 069
32. Fujiyama M (1995) Int Polym Process 10:172
33. Fujiyama M (1995) Int Polym Process 10:251
34. Fujiyama M (1996) Int Polym Process 11:159
35. Radhakrishnan S, Tapale M, Shah N, Rairkar E, Shirodkar V, Natu HP (1997) J Appl Polym Sci 64:1247
36. Sterzynski T, Calo P, Lambla M, Thomas M (1997) Polym Eng Sci 37:1917
37. Stocker W, Schumacher M, Graff S, Thierry A, Wittmann JC, Lotz B (1998) Macromolecules 31:807
38. Fujiyama M (1998) Int Polym Process 13:291
39. Fujiyama M (1998) Int Polym Process 13:406
40. Fujiyama M (1998) Int Polym Process 13:411
41. Mubarak Y, Martin PJ, Harkin-Jones E (2000) Plast Rubber Compos 29:307
42. Sterzynski T (2000) Polimery 45:786
43. Mathieu C, Thierry A, Wittmann IC, Lotz B (2002) J Polym Sci, Part B: Polym Phys 40:2504
44. Varga J, Schulek-Toth F, Pati M (1992) HU Patent 209132
45. Shi G, Cao Y, Zhang X, Hong J, Hua X (1992) Chin J Polym Sci 10:319
46. Zhang X, Shi G (1994) Thermochim Acta 235:49
47. Tjong SC, Shen JS, Li RKY (1995) Scr Metall Mater 33:503
48. Tjong SC, Shen JS, Li RKY (1996) Polym Eng Sci 36:100
49. Wolfschwenger J, deMink P, Bernreitner K (1995) DE Patent 4 420 991
50. Tjong SC, Shen JS, Li RKY (1996) Polymer 37:2309
51. Varga J, Schulek-Toth F (1996) J Therm Anal 47:941

52. Varga J, Ehrenstein (1997) Colloid Polym Sci 275:511
53. Tjong SC, Li RKY, Cheung T (1997) Polym Eng Sci 37:166
54. Tjong SC, Xu SA (1997) Polym Intern 44:95
55. Li JX, Cheung WL (1998) Polymer 39:6935
56. Varga J, Mudra I, Ehrenstein GW (1999) J Therm Anal Calorim 56:1047
57. Shi G, Zhang J, Jing H (1993) US Patent 5 231 126
58. Sterzynski T, Oysaed H (2004) Polym Eng Sci 44:2004
59. Ikeda N, Yoshimura M, Mizoguchi K, Kimura Y (1995) EP Patent 632 095
60. Davidson PMMcK, Biddiscombe HA, Govier RK, Ott MFM (1998) EP Patent 865 913
61. Konrad R, Ebner K, Bernreitner K, Wolfschwenger J (2000) EP Patent 1 044 240
62. Busse K, Kressler J, Maier RD, Scherble J (2000) Macromolecules 33:8775
63. Nezbedova E, Pospisil V, Bohaty P, Vlach B (2001) Macromol Symp 170:349
64. Kotek J, Raab M, Baldrian J, Grellmann W (2002) J Appl Polym Sci 85:117
65. Chu F, Kimura Y (1996) Polymer 37:573
66. Cho K, Nabi Saheb D, Yang H, Kang BI, Kim J, Lee SS (2003) Polymer 44:4053
67. Scudla J, Raab M, Eichhorn KJ, Strachota A (2003) Polymer 44:4655
68. Obadal M, Cermak R, Stoklasa K, Petruchova M (2003) Annu Tech Conf – Soc Plast Eng 61:1479
69. Romankiewicz A, Jurga J, Sterzynski T (2003) Macromol Symp 202:28
70. Varga J (2002) J Macromol Sci Phys 41:1121
71. McGenity PM, Hooper JJ, Paynter CD, Riley AM, Nutbeem C, Elton NJ, Adams JM (1992) Polymer 33:5215
72. Karger-Kocsis J, Varga J, Ehrenstein GW (1997) J Appl Polym Sci 64:2057
73. Karger-Kocsis J, Putnoki I, Schopf A (1997) Plastic, Rubber Compos Process Appl 26:372
74. Karger-Kocsis J, Mouzakis DE, Ehrenstein GW, Varga J (1999) J Appl Polym Sci 73:1205
75. Labour T, Vigier G, Séguéla R, Gauthier C, Orange G, Bomal Y (2001) J Polym Sci, Part B: Polym Phys 40:31
76. Tordjeman P, Robert C, Marin G, Gérard P (2001) Eur Phys J E 4:459
77. Grein C, Plummer CJG, Kausch HH, Germain Y, Béguelin P (2002) Polymer 43:3279
78. Chen HB, Karger-Kocsis J, Wu JS, Varga J (2002) Polymer 43:6505
79. Chan CM, Wu J, Li JX, Cheung YK (2002) Polymer 43:2981
80. McGoldrick J, Liedauer S, Ek CG (2002) EP Patent 1 260 545
81. Karger-Kocsis J, Varga J, Drummer D (2002) J Macromol Sci Phys 41:881
82. Bernreitner K, Hauer A, Gubo R (2003) EP 1344793
83. Gahleitner M, Hesse A, Hauer A (2004) EP Patent 1 382 638
84. Kotek J, Kelnar I, Baldrian J, Raab M (2004) Eur Polym J 40:679
85. Natta G, Corradini P (1960) Del Nuovo Cimento Suppl 15:40
86. Turner Jones A, Aizlewood AM, Beckett DR (1964) Makromol Chem 75:13
87. Immirzi A (1980) Acta Cryst B36:2378
88. Mencik Z (1972) J Makromol Sci Phys B 6:101
89. Hikosaka M, Seto T (1973) Polym J 5:111
90. Guerra G, Petraconne V, Corradini P, De Rosa C, Napolitani R, Pizozzi B, Giunchi G (1984) J Polym Sci Polym Phys 22:1029
91. De Rosa C, Napolitani R, Pizozzi B (1984) Eur Polym J 20:937
92. Napolitano R, Pirozzi B, Varriale V (1990) J Polym Sci Polym Phys 28:139
93. Norton DR, Keller A (1985) Polymer 26:704
94. Lotz B, Wittmann JC (1986) J Polym Sci Polym Phys 24:1541
95. Keith HD, Padden FJJ, Walter NM, Wickhoff HW (1959) J Appl Phys 30:1485

96. Brückner S, Meille SV, Petraconne V, Pirozzi B (1991) Prog Polym Sci 16:361
97. Meille SV, Ferro DR, Brückner S, Lovinger AJ, Padden FJ (1994) Macromolecules 27:2615
98. Dorset DL, McCourt MP, Kopp S, Schumacher M, Okihara T, Lotz B (1998) Polymer 39:6331
99. Lotz B, Wittmann JC, Lovinger AJ (1996) Polymer 37:4979
100. Cartier L, Spassky N, Lotz B (1998) Macromolecules 31:3040
101. Cartier L, Lotz B (1998) Macromolecules 31:3049
102. Zhou G, He Z, Yu J, Han Z, Shi G (1986) Makrom Chem 187:633
103. Li JX, Cheung WL, Demin J (1999) Polymer 40:1219
104. Shi G, Huang B, Zhang J (1984) Makromol Chem Rapid Commun 5:573
105. Monasse B, Haudin JM (1985) Colloid Polym Sci 263:822
106. Clark EJ, Hoffman JD (1984) Macromolecules 17:878
107. Van Krevelen DW (1997) Properties of Polymers, 3rd edn. Elsevier, Amsterdam
108. Varga J, Garzo G (1991) Acta Chim Hung 128:303
109. Varga J, Mudra I, Ehrenstein GW (1998) Annu Tech Conf – Soc Plast Eng 56:3492
110. Varga J (1994) Crystallisation, melting and supermolecular structure of isotactic polypropylene. In: Karger-Kocsis J (ed) Polypropylene, Blends and Composites, Vol. I. Chapman and Hall, London, p 56
111. Grein C, Gahleitner M, Wolfschwenger J (in preparation)
112. Kausch HH (1987) Polymer Fracture, 2nd edn. Springer, Berlin Heidelberg New York
113. Lustiger A, Markham RL (1983) Polymer 24:1647
114. Butler MF, Donald AM (1987) J Mater Sci 32:3675
115. Van der Wal A, Mulder JJ, Thijs HA, Gaymans RJ (1998) Polymer 39:5467
116. Karger-Kocsis J, Moos E, Mudra I, Varga J (1999) J Macromol Sci Phys 38:647
117. Dey SK, Agarwal PK (2001) Polyolefins 2001, Int Conf on Polyolefins, Houston, TX, 2001:441
118. Varma-Nair M, Agarwal PK (2000) J Therm Anal Calorim 59:483
119. Zhang X, Shi G (1994) Polymer 35:5067
120. Busse K, Kressler J, Maier RD, Scherble J (2000) Macromolecules 33:8775
121. Juhasz P, Varga J, Belina K (2002) J Makromol Sci Phys B41:1173
122. Fillon B, Thierry A, Wittmann JC, Lotz B (1993) J Polym Sci, Part B: Polym Phys 31:1407
123. Lotz B (1998) Polymer 39:4561
124. Sterzynski T, Lambla M, Georgi F, Thomas M (1997) Int Polym Process 7:64
125. Baran N, Stoklasa K, Pospisil L (2000) Plasty a kaucuk 5:133
126. Garbarczyk J, Paukszka D (1981) Polymer 22:562
127. Dos Santos Filho D, Oliveira CMF (1993) Makromol Chem 194:279
128. Takahashi T (2002) Sen'i Gakkaishi 58:357
129. Chen X, Wang Y, Wang X, Wu Z (1991) Int Polym Process 6:337
130. Kim S, Townsend EB (2002) Annu Tech Conf – Soc Plast Eng 60:2980
131. Fujiyama M, Kawamura Y, Wakino T, Okamoto T (1988) J Appl Polym Sci 36:985
132. Broda J (2003) J Appl Polym Sci 89:3364
133. Broda J, Wlochowicz A (2000) Eur Polym J 36:1283
134. Broda J (2004) J Appl Polym Sci 91:1413
135. Varga J (1989) J Therm Anal 35:1891
136. Scudla J, Eichhorn KJ, Raab M, Schmidt P, Jehnichen D, Haussler L (2002) Macromol Symp 184:371
137. Varga J, Breining A, Ehrenstein GW, Bodor G (1999) Int Polym Process 14:358
138. Varga J, Breining A, Ehrenstein GW, Bodor G (1999) Int Polym Sci Technol 26:20

139. Li JX, Cheung WL (1997) J Mater Process Technol 63:472
140. Labour T, Gauthier C, Séguéla R, Vigier G, Bomal Y, Orange G (2001) Polymer 42:7127
141. Roetling JA (1965) Polymer 6:311
142. Bauwens-Crowet C, Bauwens JA, Homès G (1969) J Polymer Sci A2(7):1745
143. Ward IM, Hadley DW (1993) An Introduction to the Mechanical Properties of Solid Polymers, 1st edn. Wiley, Chichester
144. Stachurski ZH (1997) Prog Polym Sci 22:407
145. Varga J, Schulek Toth F (1991) Angew Makromol Chem 188:11
146. Labour T, Ferry L, Gauthier C, Haiji P, Vigier G (1999) J Appl Polym Sci 74:195
147. Liu J, Wei X, Guo Q (1990) J Appl Polym Sci 41:2829
148. Dweik H, Al-Jabareen A, Marom G, Assouline E (2003) Int J Polym Mater 52:655
149. Lustiger A, Marzinsky CN, Mueller RR, Wagner HD (1997) 213th ACS National Meeting, San Francisco
150. Lustiger A, Marzinsky CN, Mueller RR, Wagner HD (1995) J Adhesion 53:1
151. Wagner HD, Lustiger A, Marzinsky CN, Mueller RR (1993) Compos Sci Technol 48:181
152. Tjong SC, Li RKY (1997) J Vinyl Additive Technol 3:89
153. Mi Y, Chen X, Guo Q (1997) J Appl Polym Sci 64:1267
154. Nunez AJ, Kenny JM, Reboredo MM, Aranguren MI, Marcovich NE (2002) Polym Eng Sci 42:733
155. Bhattacharya SK, Shembekar VR (1995) Macromol Reports A 32:485
156. Xie XL, Li RKY, Tjong SC, Mai YW (2002) Polym Composites 23:319
157. Grady BP, Pompeo F, Shambaugh RL, Resasco DE (2002) J Phys Chem B 106:5852
158. Jain S, Goossens H, Picchioni F, van Duin M (2003) Annu Tech Conf – Soc Plast Eng 61:1352
159. Perkins WG (1999) Polym Eng Sci 39:2445
160. Walker I, Collyer AA (1994) Rubber toughening mechanisms in polymeric materials. In: Collyer AA (ed) Rubber Toughened Engineering Plastics. Chapman & Hall, London, p29
161. Bucknall CB (2000) Deformation mechanisms in rubber-toughened polymers. In: Paul DR and Bucknall CB (eds) Polymer Blends, Vol 2. Wiley, New York p 83
162. Gaymans RJ (2000) Toughening semicrystalline thermoplastics. In: Paul DR and Bucknall CB (eds) Polymer Blends, Vol 2. Wiley, New York p 177
163. Partridge IK (1992) Rubber Toughened Polymers. In: Rostami S (ed) Multicomponent Polymer Systems. Longman Scientific & Technical Ltd., Essex, England, p 149
164. Hao WT, He YY, Luo XL, Ma DZ (2001) Chin J Polym Sci 19:317
165. Varga J, Garzo G (1989) Angew Makromol Chem 180:15
166. Varga J, Schulek-Toth F, Mudra I (1994) Macromol Symp 78:229
167. Long Y, Stachurski ZH, Shanks RA (1991) Polym Int 26:143
168. Grein C, Bernreitner K, Hauer A, Gahleitner M, Neißl W (2003) J Appl Polym Sci 87:1702
169. Grein C (2001) PhD Thesis n 2341 Swiss Federal Institute of Technology (EPFL, Lausanne, Switzerland)
170. Asano T, Fujiwara Y (1978) Polymer 19:99
171. Yoshida T, Fujiwara Y, Asano T (1983) Polymer 24:925
172. Fujiyama M (1999) Int Polym Process 14:75
173. Li JX, Cheung WL, Chan CM (1999) Polymer 40:2089
174. Li JX, Cheung WL, Chan CM (1999) Polymer 40:3641
175. Henning S, Adhikari R, Michler GH, Balta Calleja FJ, Karger-Kocsis J (in preparation)

176. Aboulfaraj M, G'Sell C, Ulrich B, Dahoun A (1995) Polymer 36:731
177. Chu F, Yamaoka T, Ide H, Kimura Y (1994) Polymer 35:3442
178. Chu F, Yamaoka T, Kimura Y (1995) Polymer 36:2523
179. Zhang X, Shi G (1994) Polymer 35:5067
180. Garbarczyk J, Sterzynski T, Paukzta D (1989) Polym Commun 30:153
181. Rybnikar FJ (1991) Macromol Sci Phys B 30:201
182. Varga J Gabor G, Ille A (1986) Angew Makromol Chem 186:171
183. Bohaty P, Vlach B, Seidler S, Koch T, Nezbedova E (2002) J Macromol Sci Phys 41:657
184. Lotti C, Correa CA, Canevarolo SV (2000) Mater Res 3(2):37
185. Ramsteiner F (1983) Kunststoffe 73:148
186. Karger-Kocsis J, Kuleznev VN (1982) Polymer 23:699
187. Grein C, Béguelin P, Plummer CJG, Kausch HH, Tézé L, Germain Y (2000) Influence of the morphology on the impact fracture behaviour of iPP/EPR blends. In: Williams JG, Pavan A (eds) Fracture of Polymers, Composites and Adhesives. ESIS Pulication 27. Elsevier Science, Kidlington (Oxford, UK), p 319
188. Read BE (1989) Polymer 30:1439
189. Boyd RH (1986) Polymer 26:323
190. Jacoby P, Bersted BH, Kissel WJ, Smith CE (1986) J Polym Sci Part B:Polym Phys 24:461
191. Grein C, Bernreitner K, Gahleitner M (accepted for J Appl Polym Sci)
192. Karger-Kocsis J, Varga J (1996) J Appl Polym Sci 62:291
193. Karger-Kocsis J (1996) Polym Eng Sci 36:203
194. Broda J (2003) J Appl Polym Sci 90:3957
195. Karger-Kocsis J (2004) DE Patent 10 237 803
196. Lustiger A, Marzinsky CN, Devorest Y (1997) US Patent 5 627 226
197. Jacoby P, Heiden M (1994) EP Patent 0 589 033
198. Kong DC, Cleckner MD (2003) US Patent 2 003 207 138, 2 002 278 241
199. Kong DC, Cleckner MD (2003) US Patent 2 003 207 137
200. Davidson PMMcK, Biddiscombe HA, Govier RK, Ott MFM (1998) EP Patent 865 914
201. Davidson PMMcK, Biddiscombe HA, Govier RK, Ott MFM (1998) EP Patent 865 912
202. Davidson PMMcK, Biddiscombe HA, Govier RK, Ott MFM (1998) EP Patent 865 911
203. Davidson PMMcK, Biddiscombe HA, Govier RK, Ott MFM (1998) EP Patent 865 910
204. Davidson PMMcK, Biddiscombe HA, Govier RK, Ott MFM (1998) EP Patent 865 909
205. Tapp WT (1992) US Patent 5 169 712
206. Jacoby P, Bauer CW, Clingman SR, Tapp WT (1992) EP Patent 492 942
207. Jacoby P, Bauer CW (1990) WO Application 9 011 321
208. Ruf BL (1996) J Plast Film Sheet 12:225
209. Fujiyama M, Kawamura Y (1988) J Appl Polym Sci 36:995
210. Hughes SK, Kody RS, Mrozinski JS, Brostrom ML (2002) WO Application 2 002 081 557
211. Ek CG, Liedauer S, McGoldrick J, Ruemer F (2003) EP Patent 1 364 986
212. Rydin C, Ek CG (2003) WO Application 2 003 087 205
213. Rydin C, McGoldrick J, Lindström T, Liedauer S (2002) EP Patent 1 260 547
214. Ek CG, Sandberg H, Liedauer S, McGoldrick J (2002) EP Patent 1 260 546
215. McGoldrick J, Ruemer F, Schiesser S, Liedauer S (2002) EP Patent 1 260 529
216. Greenshields CJ (1997) Plastic, Rubber Compos Process Appl 26:387

The Influence of Molecular Variables on Fatigue Resistance in Stress Cracking Environments

V. Altstädt

Lehrstuhl für Polymere Werkstoffe, Polymer Engineering – FAN A, Universität Bayreuth, Universitätsstrasse 30, 95447 Bayreuth, Germany
altstaedt@uni-bayreuth.de

1	**Introduction**	107
1.1	Mechanisms of Environmental Stress Cracking	109
1.2	Review of Different Conventional ESC Testing Methods	113
1.3	Fracture Mechanics Testing Methods for ESC Investigation	115
1.4	Fatigue Crack Growth Used for ESC Investigations	116
1.4.1	Important Parameters for FCP Experiments	117
1.4.2	Fatigue Crack Initiation vs. Fatigue Crack Propagation Phase	118
1.4.3	Fracture Mechanics Principles of Fatigue Crack Propagation	120
1.4.4	Fatigue Crack Propagation Diagram	122
1.4.5	Fatigue Crack Propagation at Constant Stress Intensity Factor ΔK	123
1.4.6	Factors Influencing Fatigue Resistance in Stress Cracking Environments	125
2	**Influence of Molecular Parameters on Environmental Fatigue Resistance**	126
2.1	Molecular Weight	126
2.1.1	General Trends	126
2.1.2	Amorphous Polymers	127
2.1.3	Crystalline Polymers	132
2.2	Crystallization, Chain Configuration, and Architecture	133
2.2.1	Degree of Crystallinity	133
2.2.2	Side Chain Branches	133
2.2.3	Molecular Defects vs. Crystallinity	135
2.3	Rubber Modification	136
2.4	Influence of Additives	139
3	**Multiaxial Stress Loading**	141
4	**Effect of Processing and Treatments**	143
4.1	Orientation	143
4.2	Sterilization and Aging	145
5	**Conclusion**	147
	References	148

Abstract The increasing economic importance of the polymer industry is responsible for a growing interest in the prediction of the lifetime of polymers. Although the influence of molecular parameters on the fatigue resistance in polymers has been intensively addressed, little work has been devoted to the same topic in stress cracking environments. Because of the complexity of the topic, we have studied different cases of mechanical loading, short-time, long-time, static, and dynamic in a stress cracking environment with special attention paid to fatigue, where the polymer is dynamically loaded over a long period. Fatigue crack propagation experiments can be employed as a fast and effective method for determining the long-term mechanical properties of polymers. We have particularly studied the effect of molecular weight, chain regularity, medium parameters, and processing and treatments on fatigue resistance. Special attention is also paid to the existing environmetal stress cracking prevention methods. We present an overview of the existing work and also our personal contribution to the field.

Keywords Environmental stress cracking · Fatigue crack growth · Molecular variables

Abbreviations

a_0	length of precrack
ABS	acrylonitrile butadiene styrene
ACETOPH	acetophenone
ACI	acetone
b	semiminor axis
BCN	butyl acetate
C	compliance
CED	cohesive energy density
COD	crack opening displacement
CT	compact tension
d	sample thickness (CT-specimen)
ΔE_{VAP}	heat of vaporization
ΔK	stress intensity factor amplitude
ΔK_{th}	threshold value of ΔK
ΔK_c	critical value of ΔK
dP/dx	pressure gradient
E_{dyn}	dynamic elastic modulus
EP	epoxy resin
ESC	environmetal stress cracking
ESCR	environmental stress cracking resistance
F	force
FCP	fatigue crack propagation
F_m	molar attraction constant
G_{Ic}	energy per unit area of crack
HIPS	high-impact polystyrene
IPA	isopropyl alcohol
k	Darcy's constant
K_{Ic}	critical stress intensity factor
M	molecular weight
MWD	molecular weight distribution
N	number of cycles
NMR	nuclear magnetic resonance
PA	polyamide
PC	polycarbonate
PE	polyethylene
PE-HD	high-density polyethylene
PE-LD	low-density polyethylene

PE-LLD	linear low-density polyethylene
PMMA	polymethylmethacrylate
PP	polypropylene
PS	polystyrene
PTFE	polytetrafluoroethylene
PU	polyurethane
PVC	polyvinylchloride
PVDF	polyvinylidene fluoride
R	stress ratio
S/N	stress versus number of cycles
SAN	styrene acrylonitrile
SCC	stress corrosion cracking
SEM	scanning electron microscopy
t	specimen thickness
t_{fl}	failure time in Igepal solution
T	temperature
TCB	trichlorobenzene
T_g	glass-transition temperature
U/MMA	urethane methacrylate
UHMWPE	ultra-high-molecular-weight polyethylene
VE	bisphenol-vinylester resin
V_m	molar volume
W	ligament length
X	distance along semimajor axis to point of interest
δ	Hildebrand solubility parameter
ε	elliptical rig
ε_{CR}	critical strain value
μ	liquid viscosity

1
Introduction

Environmental stress cracking (ESC) is the premature initiation of cracking and embrittlement of plastics due to the simultaneous action of stress and strain and contact with specific fluids [1]. The ESC failure phenomenon is of great practical significance for polymeric materials since it was first noted in the early 1950s. In contrast, environmental degradation of plastics is a different phenomenon, describing the change in the chemical structure of a polymer material under specific environmental conditions including thermal degradation, photodegradation, thermooxidation, hydrolysis, weathering, aging, moisture, and biodegradation [2].

If a polymer is exposed to a chemical environment, the material can undergo numerous changes. These can vary from weight gain if the polymer absorbs the chemical, to weight loss if the polymer is degraded by the chemical, or, if the chemical extracts the low-molecular-weight fraction of the polymer, to dissolution if the chemical is a good solvent of the polymer, or it can lead to other changes such as variations in color and opacity.

Environmental factors are of extreme importance to the service life of engineering components made out of plastics. The combined loading of a plastic material by mechanical stresses and physical and/or chemical active substances results in a very complex material behavior that is not identical with an exclusive mechanical loading or an exclusive physical and/or chemical attack. Therefore, the nature of the effect that the simultaneous action of mechanical and environmental parameters may have on the mechanical properties is the subject of ongoing research in the development of improved and more resistant engineering plastics. Today, it has been revealed that ESC is the most common cause of plastics product failure. ESC is responsible for more than 30% of all serious premature inservice failures of plastic components [3]. Ninety percent of these failures involve glassy amorphous thermoplastics in contact with fluids like paints, adhesives, cleaning agents, lubricants, plasticizers, inks, aerosol sprays, antirust agents, leak-detection fluids, lacquers, fruit essences, and vegetable oils. The contact is mostly unintentional or accidental. ESC might happen, for example, with pump wheels subjected to chlorine-containing flue gases or blow-molded PVC bottles exposed to some specific cleaning fluids. ESC is difficult to predict. When polyethylene (PE) pipes are used for natural gas distribution, the resistance to failure or service life is a critical consideration. The ESC phenomenon of PE, for example, was identified by Lustiger [4], and useful tools for determining the relative ESC resistance (ESCR) of various types of PE and as a quality control for assuring product quality have been developed.

The likelihood of failure due to ESC depends on the specific polymeric material, environmental conditions, and the nature and magnitude of the stress or strain applied. For economic reasons, the vast majority of data in the literature has been generated using relatively short-term tests under constant stress or strain or gradually increasing loads. The wider use of plastics in load-bearing applications has led to considerable interest in behavior under alternating loading conditions, at applied stresses that are well below the tensile yield or fracture stress. In service, many plastic components will experience fluctuating loads due to vibrating machinery or variations in pressure or temperature. For a given value of applied stress, the time to failure under cyclic loading is much less than the creep rupture time or the time to failure under steady loads. The present contribution therefore gives specific consideration to aspects of ESC behavior of plastic materials under fatigue loading conditions.

The word fatigue denotes the mechanical decay of a component subjected to variable cyclic or random forces. It is important to dinstinguish between dynamic fatigue and static fatigue. The constraint types can be very different; they provoke local deformations whose intensity and orientation can vary. This type of excitation generates heat and favors the mobility and sometimes the rupture of the molecular chains. Hence, it has the potential to be much more damaging for a polymer than a constant and high constraint. When this kind of cyclic loading is combined with an aggressive environment, the danger of polymer failure can increase dramatically: this can lead to a failure at load levels far below the limits determined under monotonous loading conditions.

1.1
Mechanisms of Environmental Stress Cracking

ESC is mostly a surface-initiated failure of multiaxially stressed polymers in contact with surface-active substances. These surface-active substances do not cause chemical degradation of the polymer, but rather accelerate the process of macroscopic brittle-crack failure. Crazing and cracking may occur when a polymer under multiaxial stresses is in contact with a medium. A combination of external and/or internal stresses in a component may be involved.

Many parameters influence the resistance of a plastic part in contact with a stress cracking environment:

- The polymer itself: chemical structure, configuration, conformation;
- Processing: molecular orientations, internal stresses, morphology, surface homogeneity;
- Environment: physical and chemical properties of medium, temperature;
- Load: type and rate of loading, velocity/duration of test;
- Geometry: size and shape of part, presence of heterogeneities or cracks.

To understand the principles of ESC, it is helpful first to look at the time dependence of the fatigue strength or the static creep rupture characteristics of an amorphous polymer tested in air. It is well known that in a long-term experiment the time to failure will increase with a decreasing level of stress or strain. For all materials, if the measurement time is long enough, a transition from a ductile yielding behavior to a brittle fracture behavior is observed. This embrittlement process may cause sudden fracture of a plastic material even at very low levels of stress. The incubation time of this naturally occurring embrittlement process is influenced by many factors, for example by the ambient temperature and the type and rate of loading, but particularly by the presence of a physical or chemical active medium. ESC is directly related to a reduction in the incubation time of stress cracking as a consequence of an attack of a physical or chemical active medium.

The decrease in the time-dependent failure strength falls into three regimes. In regime III, high load levels can be applied, and therefore the failure times are short and the influence of a stress cracking environment is only small compared to air. In the transition regime II, the time to failure is significantly reduced in the presence of a stress cracking agent. In regime I, typically very long failure times but at very low load levels are observed. In contact with different stress cracking fluids in this regime, the failure time is always significantly reduced in comparison to air.

In regime III, ductile failure is observed because at the surface of a macroscopic specimen, a high number of preexisiting microscale defects under sufficiently high stress will generate a high density of locally yielded sites, and these sites will grow and multiply with time under stress due to the time-dependent reduction in yield strength. This leads to a coalescence of the yielded sites and eventually to a macroscopic yield failure of the material. However, from regime II to regime I, when the applied stress is lowered, only a few (severe) sides will be microyielded, and these will tend to grow slowly with little chance of coalescence with near neighbors. These planar yielded zones might cavitate and fibrillate to become crazes, which eventually reach a critical length for fast unstable crack growth. Thus, high stresses in regime III promote early ductile failure and low stresses in regime I promote delayed brittle failure, while in regime II the transiton between ductile and brittle behavior occurs.

As discussed earlier, ESC of polymers is due to a more or less localized progressive degradation of the mechanical properties of the polymer starting at preexisting microscale defects. First, molecules from the environment diffuse into the polymer material and cause plasticization of the near-surface layer [5–10]. The plasticization leads to easier craze formation [6, 11, 12]. The basic mechanism of craze initiation is one of main-chain motion leading to the formation of very small voids (< 30 nm). These voids can then coalesce into planar bands that finally become crazes. Crazing is an energy-dissipating process that always happens on lower energy levels in comparison to shear yielding. It has been shown by Kramer [13] that the formation of crazes in the presence of a medium is essentialy identical to that found in air under normal conditions. Once formed, these crazes grow due to the passage of environmental molecules along the craze, leading to further plasticization of the craze tip and weakening of the craze fibrils [14]. Enhanced disentanglement and chain scission of the molecules due to a solvent is possible. Kramer [13] gives two plausible reasons why an environment can cause an accelerated craze initiation. The first reason is that the environment diffuses into the material and plasticizes it by interrupting the secondary bonds at entangled sites. This will allow easier chain motion and hence easier void formation. The plasticization hypothesis states that the solvent and vapor agents reduce the glass transition temperature and thus the yield stress near craze tips, thereby allowing flow processes to occur more readily. The second reason is that the surface energy is lowered in the presence of the liquid and so void formation is made easier. The surface energy hypothesis states that by wetting the holes in a craze, surface-active agents reduce the energy for craze formation. In a recent work, Kefalas showed that both effects complement each other rather than being mutually exclusive [15].

Besides the stress of state in the polymer, environmental stress cracking in polymers involves both solubility and absorption rate phenomena. Sensitizing media that cause ESC can be divided into two categories: those that swell or wet the polymer and those that chemically react with the polymer. The medium may be gaseous or liquid. The former mechanism has been the subject of numerous studies and is commonly recognized as the primary cause of the majority of chemically induced failures of polymers. Although both amorphous and semicrystalline polymers are susceptible to ESC, it is well known that amorphous polymers tend to be more at risk. The close packing of chains in the crystalline domains of semicrystalline polymers acts as a barrier to fluid.

A limiting factor for the growth of solvent-induced crazes in glassy homopolymers might be the hydraulic transport of the sensitizing medium through the porous craze to the craze tip [16]. If the transport is too slow, a propagating craze might be initiated by the attack of a medium, but craze growth and further fast crack propagation are not affected, given that the transport of the sensitizing medium is not fast enough to the tip of the process zone around the crack tip. This aspect might also be very important for all accelerating experiments to evaluate the ESCR of a polymer for a given application.

The driving force behind the liquid transport has been determined by Kramer and Bubeck to be capillary pressure [17]. The velocity of liquid flow V is expressed by Darcy's equation

$$V = \frac{k}{\mu}\left(\frac{dP}{dx}\right), \quad (1)$$

where μ is the liquid viscosity, k is Darcy's constant, and dP/dx is the pressure gradient.

Kambour et al. performed extensive studies on the mechanisms of plasticization [18–25]. The correlation observed between the critical strain to craze and the extent of the glass-transition temperature (T_g) depression speaks strongly in favor of a mechanism of easier chain motion and hence easier void formation. In various studies on polycarbonate [19, 24], polyphenylene oxide [20], polysulfone [21], polystyrene [22], and polyetherimide [25], Kambour and coauthors showed that the absorption of solvent and accompanying reduction in the polymer's glass-transition temperature could be correlated with a propensity for stress cracking. The experiments, performed over a wide range of polymer-solvent systems, allowed Kambour to observe that the critical strain to craze or crack was least in those systems where the polymer and the solvent had similar solubility values. The Hildebrand solubility parameter δ [26] is defined as

$$\delta = (\text{CED})^{1/2} = \left(\frac{\Delta E_{\text{vap}}}{V_m}\right), \quad (2)$$

where CED is the cohesive energy density, ΔE_{vap} the heat of vaporization, and V_m the molar volume. Hansen [10] proposed that the solubility parameter is given by contributions from the three major types of cohesive forces

$$\delta^2 = \delta_\text{d}^2 + \delta_\text{p}^2 + \delta_\text{h}^2, \tag{3}$$

where δ_d, δ_p, and δ_h are the dispersive, polar, and hydrogen bonding components of the total solubility parameter, respectively; if the solubility parameter of the solvent is close to the solubility parameter of the polymer, the polymer will most likely show some solubility as the solvent or undergo solvent-induced crystallization, in keeping with the adage that "like dissolves like." Because the solubility parameter of a polymer cannot be calculated directly from the heat of vaporization, indirect methods such as solvent swelling and group-contribution approaches are used. In the group-contribution approach [10], the solubility parameter is determined by using the equation

$$\delta = \frac{\sum F_\text{m}}{\sum V_\text{m}}, \tag{4}$$

where F_m is the molar attraction constant and V_m is the molar volume for each subsegment of the polymer repeat unit. For strongly polar and hydrogen-bonding liquids, 2D solubility parameters such as those used by Jacques and Wyzgoski [27] might have to be used.

Most of the above-cited work neglects the effect of stress or strain as a tensor. They mostly apply uniaxial stress or strain criteria. Unfortunately, most of the applications where ESC has been reported apply biaxial or multiaxial stresses to the polymer. Therefore, a more general model of the phenomenon of ESC is expected to account for generalized polymer-surface active agent systems, but also to account for generalized stress states in the material.

Kawagoe and coauthors studied ESC on polymethylmethacrylate (PMMA) and reported that ESC occurs not only under tensile stresses but also under compressive or shear stresses [28–30]. Kawagoe and Kitagawa also tried to develop a criterion for ESC, which is valid also for mutiaxial stress states in a PMMA–kerosene system. Considering pressure changes due to the solvent and T_g depression, they develop a criterion that matches their experimental data. Remarkably, this model does not take into account the thermodynamics of the system undergoing ESC at all. Kambour and others have clearly shown that the choice of a particular system that can be defined in terms of thermodynamic parameters, such as the solubility parameter, definitively affects the onset of ESC, depending on the amount of interaction between the polymer and the environmental agent. Unfortunately, their model has not been tested for any systems other than PMMA–kerosene.

A review of the recent literature related to ESC behavior and different polymer matrices is given in Table 1.

Table 1 Polymer-matrix-related literature 1990–2004 on ESC

Matrix	Reference	Matrix	Reference
PE	[31–37]	PU, PU/MMA	[8, 49–52]
PP	[31]	PET	[53]
PS	[38]	PA	[54]
SAN	[38, 39]	PVC	[10, 48, 55]
ABS	[12, 40–42]	PTFE, PVDF	[56, 57]
PC	[10, 40, 43–46]	EP	[58]
PMMA	[43, 47, 48]	VE	[59]

1.2
Review of Different Conventional ESC Testing Methods

For the development and application of a component made from a polymer material, it has to be verified that the component has a sufficient resistance against a possible attack of a corrosive medium. Obviously, the most reliable method for evaluating the stress cracking resistance of a polymer for a given application is to examine its performance under simulated end-use conditions. These experiments are time consuming and sometimes technically impossible to perform. Alternatively, environmental stress cracking resistance (ESCR) can be determined by some type of standard testing procedures whose results can be related to the stress and strain levels observed under end-use conditions. It is useful to briefly summarize here current testing methods.

A few of the numerous tests that have been developed to evaluate ESCR are listed in Table 2. The tests differ mostly in the way the external stress or strain is applied.

ASTM D 1693 describes a test for evaluating the ESCR of polyolefines in environments such as soaps, wetting agents, oils, or detergents. Strips of polymer, each containing a controlled defect, are placed in a bending rig and exposed to a stress cracking agent. The number of specimens that show cracks over a given time period is recorded.

Table 2 Commonly used tests for evaluating ESCR

ASTM D 1693	ESC of ethylene
ISO 4600	Resistance to ESC-BALL/PIN impression method
ISO 6252	Resistance to ESC-constant-tensile stress method
ISO 4599	Resistance to ESC-bent-strip method

ISO 4600 details a ball or pin impression method for determining the ESCR. In this procedure, a hole of specified diameter is drilled in the plastic. An oversized ball or pin is inserted into the hole, and the polymer is exposed to a stress cracking agent. The applied deformation, given by the diameter of the ball or pin, is constant. The test is multiaxial, relatively easy to perform, and with not very well-defined specimens, and the influence of the surface is limited. Drawbacks are the small testing surface and the undefined stress state. After exposure, tensile or flexural tests may be performed on the specimens. This leads to the determination of either the residual tensile strength or the residual deformation at break.

A constant tensile-stress method is outlined in ISO 6252, in which a test specimen is exposed to a constant tensile force while immersed in a stress cracking agent so as to determine the time to rupture under a specified stress. This uniaxial test leads to the determination of the lifetime of the specimen with accuracy, but it is time consuming and requires complex equipment. Variations of this test include a tensile creep test that monitors the strain and a monotonic creep test that uses a constant stress rate instead of a fixed stress [1].

Another bent-strip method for evaluating the ESCR is presented in ISO 4599. In this test, strips of a plastic are positioned in a fixed flexural strain state and exposed to a stress cracking agent for a predetermined period. The test is uniaxial and simple to perform, and the deformation is constant. Because of the molecular chain relaxations, the stress state is well defined only at the beginning of the test. After exposure to the medium, the strips are removed from the straining rig, examined visually for changes in appearance, and then tested for some indicative property such as tensile strength.

The critical-strain test attempts to determine the minimum strain required to initiate crazing in the presence of a stress cracking agent. The test is most commonly performed using a Bergen elliptical strain rig: a strip of plastic is placed in the rig, which is patterned like a quarter of an ellipse, and exposed to a stress cracking agent [60]. The strain at any point along the elliptical rig, ε, is given by the equation

$$\varepsilon = \frac{tb}{2a^2}\left[1 - \left(\frac{1}{a^2} - \frac{b^2}{a^4}\right)X^2\right]^{-\frac{3}{2}}, \qquad (5)$$

where a is the semimajor axis, b the semiminor axis, t the specimen thickness, and X the distance along the semimajor axis to the point of interest.

This critical-strain concept is widely used in conjunction with bend-test methods, with the aim of specifying and comparing the severity of ESC for a range of environments [11, 14, 55, 61]. It has been found experimentally that crazing occurs fairly rapidly when the strain is greater than some characteristic critical value $\varepsilon_{\text{crit}}$, while crazing remains effectively absent below this value, even for very long test times. The value of $\varepsilon_{\text{crit}}$ can therefore be used as a convenient measure of the hospitality of the environment [62]. The reason that the critical-strain concept holds for the case of bend tests is that under these conditions of the stress relaxation, the applied stress, which is the driving force for crazing, will decay relatively quickly. This is particularly true when swelling and plasticization occur as the environment diffuses into the surface and as crazes open up.

In addition to those standardized tests, two other test methods, monotonic creep and microhardness, have been developed by Hough and Wright [48]. In the monotonic creep test, the strain response to a constant stress rate is monitored. The deviation of the stress–strain characteristics in air and in the fluid of interest is taken to be the initiation of ESC. This method is shown to differentiate to a high resolution between polymers, and in the short term, the ESCR of polymer/fluid pairs that exhibit mild/weak interactions can be distinguished. The microhardness method, in which a pyramidal diamond indentor is pressed into the surface of the polymer component at a known load and for a known time, has the potential for mass screening of plastic/fluid compatibility, including extraction as well as absorption, and should be of interest to polymer suppliers.

Each of the previously mentioned tests has its advantages and disadvantages. The choice of the best method for evaluating the performance of a specific material will depend on which test most closely simulates end-use conditions (i.e., constant stress vs. constant strain, etc.) and on the failure criterion selected by the designer. Critical strain is an excellent method for evaluating stress cracking resistance if the mere appearance of crazing constitutes failure since the test determines the minimum strain required to initiate a craze. However, it should be emphasized that the appearance of crazing does not necessarily indicate a significant loss in mechanical properties of the component.

1.3
Fracture Mechanics Testing Methods for ESC Investigation

The concepts of fracture mechanics have attracted increasing interest over the last 60 years. About 20 years ago, the European Structural Integrity Society (ESIS) was asked to study the application of fracture mechanics to polymers. The ESIS TC4 testing protocol from 1990 is still the basis for characterizing the toughness of plastics in terms of the critical stress intensity factor K_{Ic} and the energy per unit area of a crack G_{Ic} at fracture initiation. Fracture mechanics is an alternative method of choice for the investigation of ESC. For fragile materials especially it offers better reproducibility than conventional test methods. In principle, the resistance of a material against crack propagation is determined. This can be done in any environment. The propagation mechanisms are very similar to the mechanisms of static failure. Several laboratories have used the concepts of fracture mechanics to evaluate stress cracking in plastics [53, 63–65].

The basic premise is that the strength of a material is determined by the presence of flaws. To study the fracture behavior, a well-defined flaw or crack is machined into a specimen prior to testing. The specimen is stressed, and the growth of the precrack in the presence of a stress cracking agent is monitored until failure. In a very simple procedure the specimen with the precrack is dipped into the test medium and loaded under constant force. The load can be applied by a simple weight or by a universal testing machine. During the experiment the increase in crack length is monitored as a function of time. If the crack starts to extend, since the load on the specimen is kept constant, an accelerated crack propagation is observed. From the dependence of the crack propagation rate on the stress intensity factor a characteristic material property as the resistance against stable crack propagation K_{Iscc} (SCC – stress corrosion cracking) can be determined and correlated with the applied medium.

1.4
Fatigue Crack Growth Used for ESC Investigations

A further development of this procedure is fatigue crack propagation (FCP) experiments in the presence of a stress cracking environment. While in the fracture mechanics test methods described above the specimen is under constant load, in a FCP experiment the specimen is tested under cyclic loading conditions in the presence of a sensitizing medium.

Besides impact, fatigue is the most critical loading mechanism for a material, especially under stress cracking environments (Fig. 1).

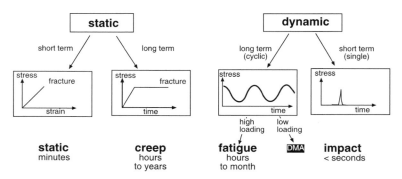

Fig. 1 Different loading conditions for polymers

1.4.1
Important Parameters for FCP Experiments

As shown in Fig. 2, the variables describing a fatigue experiment are numerous. Besides mean stress, stress amplitude, and frequency, the load type and history have to be selected in a proper way prior to testing.

For the characterization of the fatigue properties of polymers and polymer composites, Wöhler experiments to establish S–N diagrams are usually employed (Fig. 3).

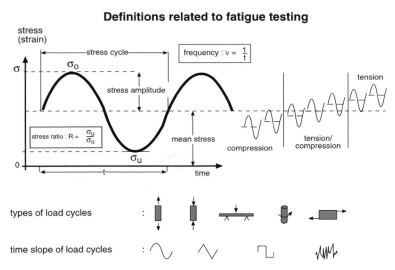

Fig. 2 Important parameters related to fatigue testing

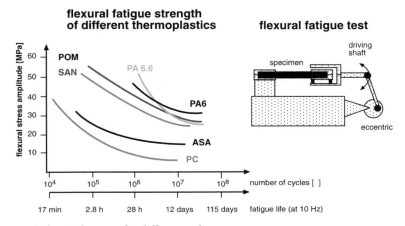

Fig. 3 Typical S–N diagram for different polymers

With this technique, the number of cycles to failure, N, is determined for different load levels, S. The Wöhler $S-N$ curve is the classical way of presenting the fatigue strength of a material. For steel there normally exists a fatigue limit below which failure will not occur since the Wöhler $S-N$ curve is practically horizontal beyond about 2 million load cycles. This, however, is conditional upon there being no influence exerted by a corrosive medium, because otherwise the Wöhler $S-N$ curve will continue to fall as the number of load cycles increases, and no fatigue limit will exist. If the specimen is completely surrounded by the stress cracking environment, $S-N$ diagrams are very suitable for monitoring the SCC behavior. But these experiments are both costly and time consuming due to the number of samples that must be stressed in an expensive test machine for days or weeks until fracture. Furthermore, the number of cycles to failure provides nearly no indication of the fatigue limits for a given application of a polymeric material because, in most cases, it is the loss of rigidity or the first irreversible material changes rather than the fracture that are of greater importance in stress cracking environments.

In contrast to Wöhler experiments, FCP experiments are a very fast and effective screening method. For a review, see Herzberg [66], Kausch [67], and Suresh and Pruitt [68–71]. The advantages of this method, such as the need for a very small quantity of test material (less than 10 g), the broad range of fatigue propagation rates measured with one specimen, and a well-defined stress state at the crack tip, favor the use of FCP experiments instead of traditional Wöhler experiments [72]. Important information about the FCP of a material in a specific environment can be obtained within 1 d by a single experiment with the help of a FCP diagram.

1.4.2
Fatigue Crack Initiation vs. Fatigue Crack Propagation Phase

Whereas fracture mechanics is ideal for studying the effect of a preexisting crack on the residual strength of a polymer material, it provides no insight into the mechanism for initiation of a crack or craze upon exposure to a stress cracking agent. Mechanical fatigue failure is due to the initiation and propagation of a crack. Both initiation and propagation might be affected by a stress cracking environment.

All fatigue failures involve two phases: the first one is the initiation of microcracks at preexisting microscale defects or other inhomogeneities that may occur at load levels far below the yield strength or the tensile strength of the material [73]. The second phase is the propagation of these cracks, which leads to total failure of the component. In order to predict the lifetime of polymers, it seems natural to explore both the initiation and the propagation phases. Specific molecular variables may have different effects on the initiation phase as compared to the propagation phase.

Therefore, it is not surprising that the tests performed on unnotched specimens happen to give significantly different results from the tests performed on notched specimens. From the fracture surface a clear distinction between thermal and mechanical breakdown is mostly possible. In unnotched specimens, significant temperature rises can occur during the loading phase even at relatively low frequencies, and these lead to a reduced fatigue durability. On the other hand, in precracked specimens, the stress is concentrated at the crack tip, allowing tests to be performed at much higher frequencies without provoking thermal failures. Moreover, a localized temperature rise in the crack-tip region can improve the fatigue performance by allowing greater plastic deformation and producing crack-tip blunting.

Figure 4 shows results investigations by Sauer and Hara [73], where the crack initiation limit for PMMA is shown in addition to the failure line. Comparing the time period from the beginning of the experiment to crack initiation, it becomes clear that the lifetime is determined by the duration of the crack propagation phase. This is the fundamental assumption of the fracture mechanical description method.

Many studies have been devoted to the influence of several factors on FCP and the total lifetime to failure; however, little information is available in the literature concerning the effect of material and test variables on craze initiation.

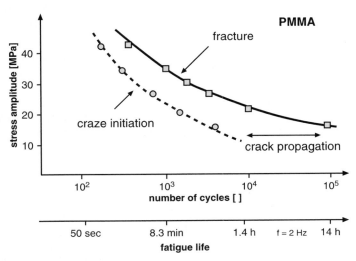

Fig. 4 Wöhlerdiagram for PMMA [73]

1.4.3
Fracture Mechanics Principles of Fatigue Crack Propagation

FCP, particularly when expressed as the fatigue crack growth rate da/dN as a function of crack-tip stress intensity factor range ΔK, characterizes a material's resistance to stable crack extension under cyclic loading. The method to study the propagation phase has been described in detail by Hertzberg and Manson 1980 [66]. Based on the ASTM E647, a new test protocol for FCP of polymers has been developed and was published in 2001 within the ESIS TC4 [74]. Each compact tension (CT) specimen is loaded dynamically with a servohydraulic test machine until fracture of the specimen. Figure 5 illustrates the CT test specimen used mostly.

The tests are usually performed at room temperature, at a frequency of 10 Hz, and the crack propagation is monitored using a front face displacement gauge. The applied waveform is sinusoidal, with a constant load amplitude, and a minimum-to-maximum load ratio, R, of 0.1. Before starting a fatigue experiment, the dynamic elastic modulus E_{dyn} of the specimen is calculated from the measured precrack length a_0, the dimensions of the CT specimen, and the specimen compliance. The initial precrack length is estimated from the crack appearance at both sides of the specimen. This value of E_{dyn} is subsequently used to calculate the crack length a as a function of N, the number of load cycles applied to the specimen. For tests under environments, the critical fluid can be applied through a soaked sponge fixed on both sides of the specimen as an unlimited source.

In each test, a precrack is first introduced ahead of the machined notch using a razor blade in order to minimize a possible effect of the plastic deformation ahead of the precrack. This initial precrack is then extended by approximately 2 mm under computer control by increasing the load ampli-

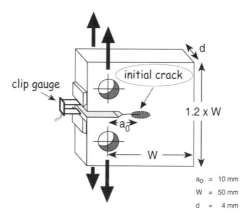

Fig. 5 Compact tension (CT) specimen

tude until a crack growth rate of 0.1 μm per cycle is obtained. For the first specimen of a given material, the load amplitude is then reduced until fatigue crack growth is no longer detectable, thereby defining ΔK_{th}, the nominal threshold value of the cyclic stress intensity amplitude. Afterwards, the same specimen is tested with an increasing ΔK: the load amplitude is raised slightly so that the crack propagates at a gradually accelerating rate, because of the increase in ΔK with crack length. This leads to the definition of ΔK_c, the critical value of cyclic stress intensity amplitude. During loading, an increase in crack length resulting from fatigue crack propagation is observed indirectly through a quasicontinuous measurement of the compliance C of the sample as a function of the number of cycles. The change in crack opening displacement (COD) is measured at the front of the CT specimen with a COD gauge. At the same time, the force amplitude $F_{max} - F_{min}$ is measured. From ΔCOD and ΔF the change in compliance C can be calculated. The test frequency f and the stress ratio $R = F_{max}/F_{min}$ are held constant. Usually, tests are only conducted in the opening mode (mode I). Depending on the software of the controller, the experiments might also be conducted under ΔK control. The crack length a is calculated by a compliance technique published by Saxena and Hudak [72]:

$$\frac{a}{W} = 1.0002 - 4.0632u + 11.242u^2 - 106.04u^3 + 464.33u^4 - 650.68u^5. \qquad (6)$$

The quantity u is given by

$$u = \frac{1}{\sqrt{C E_{dyn} d} + 1}, \qquad C = \frac{COD_{max} - COD_{min}}{F_{max} - F_{min}}, \qquad (7)$$

where W is the ligament length, d is the specimen thickness, E_{dyn} is the elastic modulus, C is the compliance, F is the load, and COD is the crack opening displacement.

The amplitude of the stress intensity factor ΔK for mode I loading of a compact tension specimen is given as

$$\Delta K = f\left(\frac{a}{W}\right) \times \left(\frac{F_{max} - F_{min}}{d\sqrt{W}}\right), \qquad (8)$$

where

$$f\left(\frac{a}{W}\right) = \left(\frac{2 + \frac{a}{W}}{\left(1 - \frac{a}{W}\right)}\right) \qquad (9)$$

$$\times \left[0.886 + 4.64\left(\frac{a}{W}\right) - 13.32\left(\frac{a}{W}\right)^2 + 14.72\left(\frac{a}{W}\right)^3 - 5.6\left(\frac{a}{W}\right)^4\right].$$

1.4.4
Fatigue Crack Propagation Diagram

In a FCP diagram, the crack propagation rate da/dN is illustrated as a function of the amplitude of the stress intensity factor ΔK:

$$\Delta K = K_{\max} - K_{\min}. \tag{10}$$

Paris [75] showed in 1964 that a linear relationship predicted by a simple power law for a double-logarithmic scale exists between the FCP rate da/dN and the applied ΔK (Fig. 6). The linear dependence is frequently observed only over an intermediate range of growth rates. On the double-logarithmic scale, the region of stable crack growth is usually satisfactorily described by the Paris–Erdogan equation

$$da/dN = m_0(\Delta K)^m, \tag{11}$$

where the parameters m_0 and n are material constants and ΔK is the stress intensity factor range at the tip of the crack. In general, the exponent n in Eq. 11 varies with the R-ratio, but it frequently has a value close to 4. When investigating a wide range of crack propagation rates, deviations from this linear behavior may be observed, as illustrated schematically in Fig. 6. That is, FCP rates decrease rapidly to vanishingly small values as ΔK approaches the threshold value ΔK_{th}. This ΔK level defines a design criterion that is analogous to the fatigue limit determined from traditional S–N curves [66]. FCP rates increase markedly as ΔK approaches K_{cf}, at which unstable fracture occurs within one loading cycle. From the viewpoint of evaluating a materials fatigue resistance, any decrease in FCP rates at a given value of ΔK or, alter-

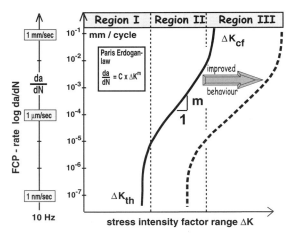

Fig. 6 FCP diagram

natively, any increase in ΔK to drive a crack at a given speed is, of course, beneficial.

1.4.5
Fatigue Crack Propagation at Constant Stress Intensity Factor ΔK

Obtaining a distinct curve of da/dN vs. ΔK requires a well-defined crack propagation rate at each single value of ΔK. During crack propagation, the crack tip is passing through dissimilar volume elements of the specimen that are assumed to be similar in properties. Because any material constrains a certain amount of inhomogeneities, there exists a scatter in the local crack propagation rate that reflects these inhomogeneities in the material (assuming signals from the load cell and the COD transducer with very low noise). To investigate the degree of homogeneity of a material, with a suitable software it is possible to control the FCP through the specimen at constant ΔK. At a constant ΔK, the crack propagation rate should be constant in a homogeneous material. Thus the scatter in the crack propagation rate da/dN at constant ΔK test conditions reflects the amount of heterogeneity in the material. By this approach it is also possible to generate a fracture surface at a constant crack propagation rate da/dN of 1 or 15 mm in length, which reflects the resulting morphology at a predetermined crack propagation rate.

As an example, Fig. 7 shows a plot of da/dN vs. crack length ratio a/W (which is proportional to the actual crack length) for different constant ΔK

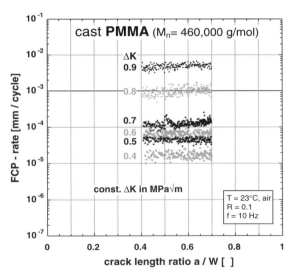

Fig. 7 Plot of da/dN vs. crack propagation rate for W 33 CT specimens from cast PMMA at different constant ΔK levels

values. Each curve reflects the behavior of one single specimen. For the CT specimens used in this experiment, an a/W ratio of 0.4 corresponds to a crack length of approx. 13.2 mm and a ratio of 0.7 corresponds to a length of approx. 23.1 mm. The crack was extended by about 10 mm at each ΔK level. For all curves, the ordinate on a logarithmic scale shows a typical, but small, scatter. The graph of these data vs. ΔK is shown in Fig. 8 together with an ordinary FCP curve of the same material measured with a delta ΔK increase of 0.08 MPa $\sqrt{m}$. The length and position of the vertical bars measured at various ΔK levels and with different CT specimens agrees well with the curve which determined with one single CT specimen in a ΔK increase experiment. Thus, performing FCP measurements at an appropriate constant stress intensity factor ΔK is an excellent means to demonstrate the homogeneity of a material and to identify the prevalent micromechanical mechanism of FCP from the corresponding fracture surfaces.

Another important challenging possibility obtained by constant ΔK-FCP experiments is direct investigations of a propagating crack exposed to a chemical environment. Assuming an initially constant crack propagation rate, the injection of a medium that will penetrate to the crack tip by capillary forces can accelerate or decelerate crack propagation, depending on the micromechanical effect of the medium on the polymer at the crack tip. This will induce time-dependent changes in the crack propagation. Figure 9 depicts a plot of a/W vs. N for an inhomogeneous crack propagation. During the first 10 000 cycles the crack propagates at a constant rate represented by the constant slope in Fig. 9. At point A, isopropyl alcohol (IPA) was injected. After the

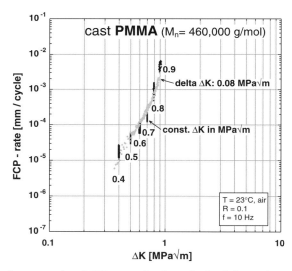

Fig. 8 Buildup of a conventional FCP curve by data obtained from Fig. 7

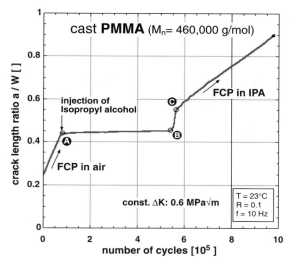

Fig. 9 Plot of crack length ratio vs. number of cycles provides information about time-dependent inhomogeneous crack propagation

injection, between points A and B, very slow crack propagation was observed. After 55 000 cycles (point B), the crack was observed to suddenly speed up until point C, where stable crack propagation was again observed, although at a lower rate as in air. Explanations for this manner of crack growth behavior can be found in detail in the following chapters.

1.4.6
Factors Influencing Fatigue Resistance in Stress Cracking Environments

The following sections give consideration to various aspects of environmental stress cracking in polymers and particularly to fatigue crack growth of polymers in contact with a stress cracking agent. Three groups of intrinsic and extrinsic parameters have a noticeable influence on the lifetime of a specimen. These three groups are linked to the material itself, to the specimen, and to the testing conditions. For the material itself, in the case of polymers, the chemical structure and cross linking are important, but one also has to consider the molecular weight and molecular weight distribution (MWD), molecular orientation, and crystallinity. Semicrystalline polymers with a high molecular weight M_w and a narrow distribution M_w/M_n are the most qualified for fatigue resistance. Furthermore, all factors that are susceptible to reduce initiation and propagation of crazes will improve the lifetime in a specific environment.

The geometry of a plastic component is particularly important: an appropriate design will eliminate stress concentrations and adequate processing conditions will lead to a flaw-free surface possibly coupled with a postprocessing treatment. Concerning the loading conditions, the nature and the viscosity of the medium are important parameters. One should prevent or eliminate the possibility of a chemical attack of the polymer by the medium and pay special attention to those media where the solubility parameters are close to the parameters of the polymer.

We have chosen to focus on the molecular variables that influence fatigue resistance in a stress cracking environment: molecular weight, chain regularity, and molecular parameters of the medium. In most cases, we will differentiate between amorphous, crystalline, and cross-linked polymers. In a subsequent section we will examine the impact of sample preparation on the fatigue resistance: sterilization, cross linking, orientation. Another section will focus on the different strategies to improve the ESCR.

Although a great amount of work has been devoted to the analysis of the influence of molecular variables on the fatigue resistance of polymers, little work has addressed the influence of molecular variables in stress cracking environments.

2
Influence of Molecular Parameters on Environmental Fatigue Resistance

2.1
Molecular Weight

2.1.1
General Trends

Various studies have been performed to determine the influence of molecular weight on fatigue resistance of polymers under cyclic loads. In general, physical properties of polymers are strongly dependent on the average molecular weight M_w and the width of the MWD, M_W/M_N, since the presence of molecular entanglements can significantly affect the mechanical behavior. A higher molecular weight means higher environmental stress crack resistance. A broadening of the MWD, especially in the medium molecular weight range, decreases the ESCR because the weak areas produced by low-molecular-weight fractions cannot be compensated.

For fatigue resistance in air, there is a noticeable increase in the resistance for smooth, unnotched specimens and a significant increase in the fatigue resistance for prenotched specimens as the molecular weight is increased. The fatigue resistance of polymers in a stress cracking environment is also generally increased as the molecular weight increases. This will be shown for different polymers in this section.

2.1.2
Amorphous Polymers

Amorphous polymers are especially sensitive to stress cracking. Therefore, polystyrene (PS) appears to be an appropriate material for the study of the corresponding effects.

To determine the influence of the molecular weight on the ESC behavior under fatigue crack growth conditions, two commercial PS polymers were selected for an investigation by Altstädt et al. [76], see Table 3, under air and aggressive medium. The selected fluid was a commercial sunflower oil (tradename Livio). The tests under oil were carried out by continuously applying the fluid through a soaked sponge that was fixed on both sides of the CT specimen. With this approach both sides of the crack tip were always covered with the soaked sponge.

The molecular weight of PS 148H is $M_w = 238\,000$, whereas PS 168N has a notably higher value of $M_w = 354\,000$. As shown in Figs. 10 and 11, both PS systems exhibit a severe decrease in fatigue crack growth resistance when tested under the influence of the oil: The whole curve is shifted toward lower ΔK values, which implies a decrease in the threshold value ΔK_{th}, and there is an increase in the slope of the curve.

The comparison of both PS systems clearly shows an improved ESCR under fatigue loading of the PS with an increasing molecular weight, both with and without oil. This is reflected in a lower slope of the FCP curve and its displacement to higher ΔK values such as ΔK_c and ΔK_{th} for PS 168N.

FCP experiments for PMMA types of different molecular weight have also been performed in air and IPA using the sponge method described above. The materials tested were two injection molding grades with $M_N = 45\,200$ and 67 200, ($M_W/M_N = 2.1$). They were compared to a cast PMMA with $M_N = 460\,000$ ($M_W/M_N = 5.6$). As inhomogeneous crack propagation was expected for the fatigue in alcohol, the constant ΔK method was applied for testing. Figure 12 left shows the three PMMA grades tested at a constant ΔK of 0.6 MPa $\sqrt{m}$. The ranking of the polymers on the logarithmic scale corresponds to the molecular weight, with the PMMA with $M_N = 460\,000$ showing the lowest crack propagation rate.

Table 3 Investigated materials

Material	Property
PS 148H®	PS – M_w 238 000 g/mol
PS 168N®	PS – M_w 354 000 g/mol
PS 486M®	PS impact modified by small PB particles
PS 2710®	PS impact modified by large PB particles

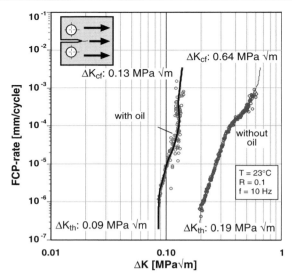

Fig. 10 FCP diagram of lower-molecular-weight PS with and without stress cracking medium (PS 148H)

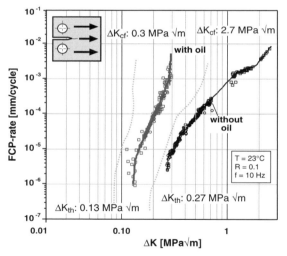

Fig. 11 FCP diagram of higher-molecular-weight PS with and without stress cracking medium (PS 168N)

Figure 12 right illustrates the effect of IPA. The experiments were first performed in air, as a reference, up to a crack length ratio of about 0.45; then the alcohol was added by the sponge without interruption of the experiment. Under the influence of the medium, the crack propagation rates changed instantly. The highest increase of about two decades accompanied

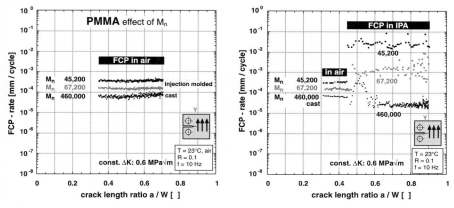

Fig. 12 Plots of da/dN vs. crack length ratio at ΔK of 0.6 MPa $\sqrt{m}$ for three PMMA types of different molecular weight. *Left*: testing in air, *right*: testing in IPA

by a pronounced scatter in the propagation rate was found for the injection molding grade with $M_N = 45\,200$. The increase for $M_N = 67\,200$ was less than one decade, demonstrating the strong influence of the molecular weight on the ESCR. In contrast to the behavior of the lower M_N PMMAs, the cast PMMA of $M_N = 460\,000$ shows at first very unstable crack propagation (up to an a/W of 0.6); the rate stabilizes thereafter at a level significantly lower than that in air. This observation is quite remarkable and can be well explained by considering the present ideas on craze formation and fibril stability as outlined in the contribution of Monnerie et al. in this volume. The fact that the fracture surface of a fatigue crack in air of PMMA $M_N = 67\,200$ shows no visible features indicates that the craze fibrils break, mostly due to disentanglement, almost without affecting the matrix. This explains correctly the M_w effect on the rate of air fatigue cracks in PMMA.

With IPA added to the lower M_w PMMAs, two effects are observed: craze fibrils disentangle more rapidly and the plasticized matrix deforms plastically leaving striations on the fracture surfaces (Fig. 13). The high M_N PMMA behaves differently; we can assume that fibril failure in air involves quite a bit of chain scission as found by Monnerie et al. (this volume) in low-temperature failure of amorphous polyamides. With the addition of IPA chain scission is reduced, and fibrils will strain harden through orientation, thus increasing the plastic deformation of the matrix and the resistance to (fatigue) crack propagation as compared to the behavior in air. The striations on the fracture surface (Fig. 13) are typically associated with discontinuous crack propagation. Here, a craze is developing during a large number of cycles without any measurable crack propagation before a stepwise crack propagation occurs. This phenomenon was first identified with PVC and polycarbonate (PC) after an analysis of the fracture surface showed that the number of markings did not agree with the number of load cycles. Therefore, an analysis of the frac-

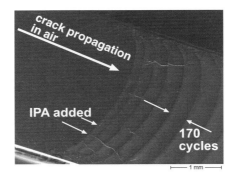

Fig. 13 A SEM micrograph of striations in a PMMA sample with $M_N = 67\,200$ tested at $\Delta K = 0.6$ MPa $\sqrt{\text{m}}$ in IPA

ture surface with the SEM should always be accompanied by fatigue testing. As an example, in Fig. 13 with PMMA tested in IPA the distance between two markings was found to be 170 load cycles.

The effect of the MWD on polymer fatigue is illustrated by experiments on two PS grades with similar molar mass [79]. The GPC curves in Fig. 14a of the two polymers with similar weight-average molar masses show a low polydispersity index M_W/M_N of 1.1 for polymer A, which resembles a narrow distribution of chain length, whereas polymer B has a broad range of $M_W/M_N = 2.71$ (Fig. 14a). As can been seen in Fig. 14b, for the material with the lower M_W/M_N ratio the whole FCP curve is shifted toward higher ΔK values; the crack propagation rate in region II of the $da/dN - \Delta K$ diagram is especially reduced.

In comparison to this behavior, Fig. 14c depicts FCP curves of PS materials with different MWDs but a similar number-average weight of $M_N = 90\,000$. Although M_W is more than twice as high for polymer A compared to polymer B, the latter has superior fatigue properties.

Both examples show that for wide MWDs a detrimental effect on the fatigue behavior is caused by the low molecular weight fraction, which cannot build up effective entanglements. This effect may even overcome the influence of a higher weight-average molecular weight. In fact, experiments in stress cracking media have not yet been performed, but the demonstrated influence would be expected to be valid for those conditions as well.

Moskala [53] discovered that the stress required to initiate a craze is independent of the polymer molecular weight but that the stress required to propagate a craze increases with increasing molecular weight. This result suggests that the ability of a polymer to arrest a growing crack and to form a stable craze depends on its molecular weight. Several amorphous polyethylene terephtalate types of different inherent viscosities, i.e., molecular weights, showed the same threshold value ΔK_{th} but a rising ΔK_C for increasing vis-

The Influence of Molecular Variables on Fatigue Resistance

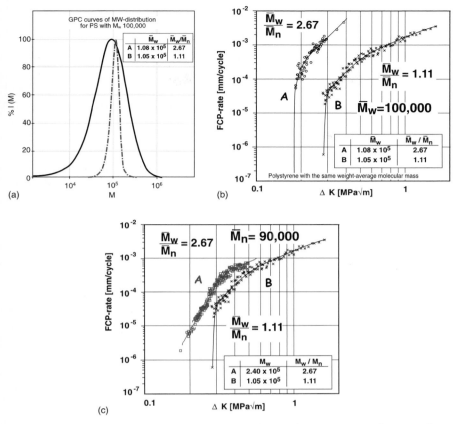

Fig. 14 a GPC curves of MWD for two PS types with M_w of 100 000. **b** FCP diagram of two PS materials with similar weight-average molecular weight of 100 000 but different weight distributions (Fig. 14a). **c** FCP diagram of two PS materials with similar number-average molecular weight of 90 000 but different weight distributions

cosities when tested in constant load experiments on CT samples in aqueous sodium hydroxide.

An improvement in ESCR with increasing molecular weight in n-butanol has been shown by Rudd [80] for PS, where the logarithm of the time to fracture increases proportionally with the quantity M_w. Wellinghoff and Baer [81] have shown that the fibril volume fraction of PS crazes increased with increasing molecular weight, which can contribute to a greater resistance to liquid transport. In addition, an increased M_w results in a longer craze fibril breakdown time, as shown by Kramer and Bubeck [17].

2.1.3
Crystalline Polymers

The fatigue resistance of partly crystalline polymers is supposed to depend predominantly on the size and distribution of the crystalline domains. The interfaces between the different individual components, e.g., lamellea/lamellae, spherulite/spherulite, and lamellea or spherulite and amorphous phase are responsible for weak points in the polymer. So-called tie molecules can be considered to act as a binder between different lamellae, with a binding strength proportional to the valence forces. Separation between adjacent lamellae, either by disentanglement or rupture of the tie molecules, is supposed to be the main molecular mechanism for ESC. Due to the improved binding strength of the chains, a higher molecular weight also favors the resistance to ESC.

In the following discussion, results from ESC testing by means of diverse methods are described.

Using static fracture mechanics measurements Rufke showed that the storage of high-desnity polyethylene (PE–HD) of different molecular weights in a 5% solution of dispergator (alkyl phenol polyglycole ether) leads to a different behavior with respect to stress cracking [82]. As seen in Fig. 15, the material with the higher molecular weight shows a higher threshold value and a lower slope of the curve, leading to a significantly slower growth of the induced crack.

For crystalline polymers, however, not only the molecular weight and polydispersity index are of interest. Besides the degree of crystallinity, there is a specific influence of chain design, i.e., existence, length, and concentration of side chains on the resistance against environmental stress cracking, as will be pointed out in the next section.

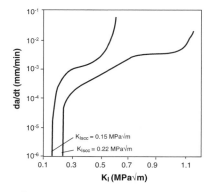

Fig. 15 Dependence of crack propagation rate da/dt on DK for two PE–HD materials of different molecular weight in a 5% dispergator dispersion [82]

2.2
Crystallization, Chain Configuration, and Architecture

2.2.1
Degree of Crystallinity

In general, crystalline polymers show a higher resistance to ESC than comparable amorphous ones, as could be demonstrated for different types of PS [83]. For a syndiotactic crystalline material with a molecular weight of only $M_W = 190 000$ undergoing a long-term tensile test in methanol, the stress at fracture was markedly higher than that of a high-molecular-weight atactic amorphous PS ($M_W = 325 000$). In crystalline polymers, tie molecules that are anchored in the lamellae hinder the disentanglement, which generally implies that the higher the degree of crystallinity, the better the resistance to ESC.

For PE, the so-called Igepal transition time, which is the time during an experiment in which the medium becomes effective, was found to increase significantly with a rising degree of crystallinity when the molecular weight and weight distribution as well as the branch frequency were kept similar [35].

2.2.2
Side Chain Branches

Besides crystallinity itself, the kind and amount of side chains is of importance for crazing. A schematic investigation of the influence of chain branches was performed in PE that were introduced via copolymerization [84]. A long-chain branched low-density polyethylene (PE–LD) was compared to a linear low-density polyethylene (PE–LLD) with different short chains [85]. Type and concentration of the copolymers were chosen in order to attain the same density of 0.920 g/cm^3 and melt flow rate of $25 \text{ g}/10 \text{ min}$ for all polymers. The ESC resistance was measured in a long-term tensile test of notched specimens at 50 °C in 10% Igepal solution.

Table 4 shows the failure time t_{fl}, which increases significantly with concentration and length of the comparatively short (C3–C8) branches of PE–LLD, which is due to the increased sliding resistance of the polymer chains through the crystal. It is well known that methyl and, to a lesser extent, ethyl side branches can become part of the crystalline lamellae of PE. Increasing the size of these short-chain branches decreases the cocrystallization probability and, therefore, favors the formation of tie molecules. The lowest fracture time, i.e., the worst performance, is found for the long-branched PE–LD, which substantiates the notion that long chains are not efficient barriers against disentanglement.

In accordance with these results, ESCR as the failure time of PE in Igepal, measured by static notch opening displacement, increases dramatically as the short-chain length increases from 2 to 4 and 6 carbon lengths. The tests

Table 4 Time to fracture of branched PEs tested in long-term tensile tests at 50 °C in a 10% aqueous solution of Igepal (CO-630, notched specimen). Number of branches/1000 carbon atoms: 18. Isothermal crystallization at 115 °C/1 h [85]

Sample	Comonomer	M_w	MWD	Branch typ	t_fl at 9 MPa
A	1-Butene	9.2×10^4	3.4	Ethyl	570
B	1-Hexene	9.4×10^4	3.3	Butyl	860
C	1-Octene	9.7×10^4	3.6	Hexyl	6200

were performed with samples of similar molecular weight and weight distribution as well as similar branch frequency but different short-chain branch lengths [35].

The effect of both crystallization and branching was studied for 19 different PE types produced by different polymerization processes and catalyst types, which were characterized by the so-called crystallization analysis fractionation. The solubility of PE fractions in trichlorobenzene (TCB) refers to their ability to create tie molecules and therefore correlates with ESCR, which was measured with the standard bent-strip method in Igepal CO-630 [84]. The fractions were distinguished by their solubility in different temperature ranges, where the solubility depends on either crystallinity and branching:

- The fraction of polymer chains soluble above 85 °C is highly crystalline and therefore has very few short-chain branches and contains only a few tie molecules.
- Tie molecules are most likely present in the fractions that dissolve between 75 and 80 °C. Corresponding chains are part of the polymer crystalline matrix but have some chain segments containing short branches that will not crystallize but link the crystalline sections.
- Polymer chains that are soluble below 75 °C are not able to crystallize sufficiently to form very effective tie molecules.

It was found that branches in high-molecular-weight types of linear ethylene copolymers were more effective in withstanding ESC than those in low-molecular-weight polymers [85]. The authors synthesized groups of PE polymers of concordant molecular weight, i.e., inherent viscosity, with increasing concentration of ethyl groups. Figure 16 illustrates six curves produced by the ESC results of these groups in a bending test called ORL in a 10% aqueous solution of a surface active agent (Nonion). The curves show modest slopes for low-molecular-weight polymers and high increases for high-molecular-weight types. This shows that the latter have a higher potential to act as tie molecules, leading to an enhanced performance under environmental influences.

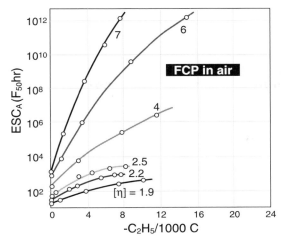

Fig. 16 ESC (50% failure) of various ethylene-butene copolymers for different amounts of C_2H_5 branches and different molecular weights (inherent viscosities)

Nevertheless, it should also be noted that high degrees of branching lead to a reduced ability to crystallize and can therefore reduce ESCR.

2.2.3
Molecular Defects vs. Crystallinity

A counterexample for the role of crystallinity in ESCR can be found for polyvinylidene fluoride (PVDF, $[CH_2CF_2]_n$) [86]. This high-molecular-weight, semicrystalline fluorine polymer is widely used in corrosion-resistant applications where a high degree of crystallinity is often the key to resistance. As PVDF suffers from a remarkable lack of stability in basic environments due to dehydrofluorination, commercial homopolymers with a different degree of crystallinity have been evaluated in 30% sodium hydroxide using 5% and 7% of constant strain in a bending test.

It was found that the ESCR decreased with an increasing degree of crystallinity. This behavior has been related to the presence of "head-to-head (CF_2-CF_2)" and "tail-to-tail (CH_2-CH_2)" sequences, also called "HHTT-regiodefects". They inhibit the basic degradation mechanism; moreover, they decrease chain regularity and hence the degree of crystallinity. The content of the regiodefects was evaluated by ^{19}F NMR analysis and is proportional to the measured inversion. Figure 17 shows the time to crack appearance vs. the NMR inversion. The less-crystalline PVDF polymer with the highest content of the regiodefects presents the best crazing resistance in sodium hydroxide for both strains.

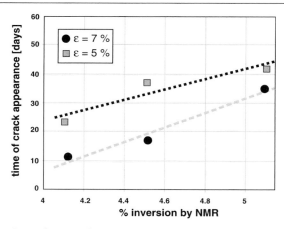

Fig. 17 Cracking time of PVDF sheets vs. percent inversion as determind by NMR, i.e., content of regiodefects

2.3
Rubber Modification

A common strategy for improving the mechanical behavior of thermoplastics is to incorporate uniformly distributed rubber particles in the polymer matrix. For high-impact polystyrene (HIPS) styrene-butadiene block copolymers are used. In order to investigate whether rubber toughening also enhances ESCR, two commercial PS grades with a different average diameter of rubber particles were selected (PS 486M: $d = 2\,\mu m$, PS 2710: $d = 5\,\mu m$) [76]. The PS materials were tested both with and without exposure to vegetable oil. The corresponding FCP diagrams are shown in Fig. 18a,b.

The FCP behavior of the two rubber-modified polystyrenes is comparable in the threshold region and in the region of unstable crack growth (region III) at high propagation rates. A significant difference is visible in region II of intermediate propagation rates 10^{-5} to 10^{-4} mm/cycle.

Compared to the unmodified PS of different molecular weight (PS 148H and PS 168N), shown in Figs. 10 and 11, the observed ranking in terms of FCP behavior without media is PS 148H < PS 486M = PS 2710 < PS 168N. This ranking is completely changed when the tests are conducted in the presence of vegetable oil: PS 148H $\ll$ PS 168N < PS 486M < PS 2710. Obviously, under ESC conditions, rubber toughening is more efficient for fatigue resistance than an increase in molecular weight. A remarkable difference between the two rubber-modified systems is visible in the threshold region. The ΔK_{th} value of PS 2710 is not affected by the presence of the medium, while ΔK_{th} of PS 486M is reduced by a factor of two.

Analysis of the fracture surfaces in Fig. 18c reveals the different size of the rubber particles in the two PS systems. Tested in air, in both cases the crack

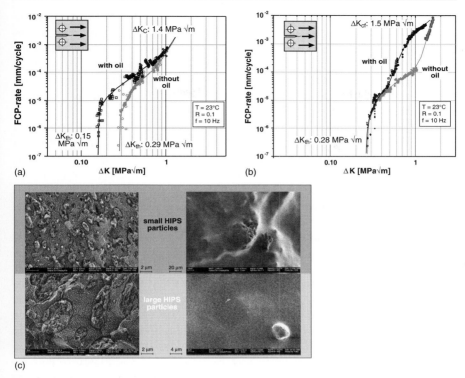

Fig. 18 a Effect of small styrene-butadiene rubber particles incorporated in PS (PS 486M) on FCP with and without oil. **b** Effect of large styrene-butadiene rubber particles incorporated in PS (PS 2710) on FCP with and without oil. **c** SEM micrographs of PS fracture surfaces with smaller rubber particles (*above*) and larger rubber particles (*right*). *Left*: without oil, *right*: with oil

propagates through the rubber particles, which is possible only in the case of good interfacial bonding. Tested in vegetable oil, the fracture surface appears to be very smooth. In the case of PS 486M, some marks from the underlying morphology are visible.

It is generally accepted that rubber particles act as stress concentrators, initiate crazes, and participate in their termination. The contribution to the toughness depends on the concentration of the rubber, the particle size, and the interfacial bonding. In the case of fatigue crack growth small particles are more efficient at low crack propagation rates because the plastic zone is smaller, while larger particles are more efficient at higher propagation rates and the corresponding larger plastic zone diameters.

Under the influence of a stress cracking medium it appears that in the case of large rubber particles small crazes may easily develop. As a consequence, the diffusion rate of oil is reduced because of the increased fibril density.

Small rubber particles show less intensive crazing, and the crazes are more easily overloaded because of a higher effective opening of the crack tip. In the latter case the oil can penetrate through the crazes more easily. This effect is stronger in region I of the FCP diagram than in region II. At high crack propagation rates, it seems possible that the oil may be hindered from penetrating fast enough into the crack tip, which explains the similar ΔK_c for both rubber-modified systems.

The mutual influence of rubber size and molecular weight was studied using a series of nine HIPS samples with 8% rubber content differing in rubber particle size (2, 4, and 8 μm) and average molecular weight (M_w 220 000, 245 000, and 270 000) [87]. The compression-molded samples containing a semicircular stress concentrator were stressed with a constant load of 5.5 MPa being in contact with a simulated food oil (50/50 solution of cotton seed oil and oleic acid). As shown in Fig. 19, increasing the molecular weight as well as the rubber particle size has a positive influence on ESCR— the highest molecular weight, coupled with the biggest particles, proves to be the best.

It is well known that rubber modification of PMMA improves its fracture toughness. With constant ΔK measurements it is possible to investigate the influence of a medium, in this case IPA, on the FCP behavior in the presence of rubber particles. In Fig. 20 a rubber-modified commercial PMMA grade with a molecular weight M_N of 45 200 is compared to the previously discussed nonmodified polymers with molecular weights M_N of 45 200 and 67 200. As expected, tested in air, the rubber-modified PMMA shows a significantly lower crack propagation rate in comparison to the unmodified polymer. Remarkably, the crack propagation rate is nearly one decade lower in air as

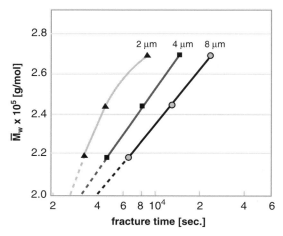

Fig. 19 ESCR in 50% cotton oil 50% oleic acid for 8% rubber content high-impact PSs as a function of weight average molecular weight and gel particle size (2, 4, and 8 μm)

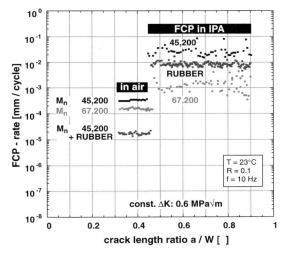

Fig. 20 Rubber-modified PMMA of $M_N = 45\,200$ compared to unmodified PMMA with molecular weight $M_N = 45\,200$ and $67\,200$. FCP rate vs. crack length ratio a/W, performed in air (up to $a/W = 0.4$), and IPA

compared to the PMMA with the higher molecular weight ($M_N = 67\,200$). In the presence of the medium IPA, the ranking of the polymers with regard to the crack propagation rate has changed significantly. The rubber-modified PMMA performs with a lower FCP rate compared to the unmodified PMMA, but the polymer with the higher molecular weight ($M_N = 67\,200$) exhibits a nearly one decade lower crack propagation rate. This clearly proves the importance of molecular entanglements in resisting ESC.

2.4
Influence of Additives

The processes at the crack tip of a notched specimen in a stress cracking liquid under mechanical stresses and possibly elevated temperatures can be regarded as local aging phenomena. Fracture mechanics, on the one hand, is a very sensitive method of detecting a polymer's behavior with regard to these influences. On the other hand, as demonstrated in the previous examples, FCP measurements can distinguish between small material differences. As depicted in Fig. 21, PE–HD shows the different behavior in static fracture mechanics tests with respect to water at a temperature of 80 °C for two different stabilizers. Stabilizer 2 leads to higher K values at comparable da/dt values and is therefore more appropriate for hot-water applications when oxidation processes also have to be taken into account [88].

For medical devices, in particular for components of infusion systems, transparent polymers are indispensable. Due to the multitude of drugs possibly flowing through these systems, the need for disinfection of the devices and the possibility of connecting various components by force locking, the ESCR plays a decisive role in the selection of a suitable material [76]. Figure 22a shows an example of a three-way stop cock as an important component of such a system.

In principle, ESCR can be controlled within limits that are mostly given by the viscosity for processing, by the molecular weight, and by the MWD of the polymer. For these reasons polycarbonate (PC) with a molecular weight of more than 30 000 g/mol is used for this application. In the case of par-

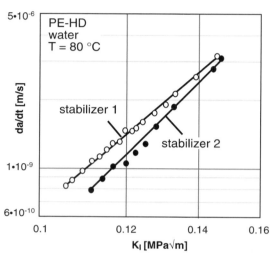

Fig. 21 Influence of type of stabilizer on crack propagation rate in PE–HD in water at $T = 80\,°C$; static fracture mechanics

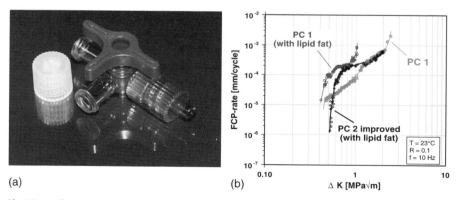

Fig. 22 **a** Three-way stop cock as a component for medical infusion system. **b** FCP plots of PC with $M_W = 30\,000$: an improvement in material properties leads to better fatigue resistance of PC2 in lipid fat emulsion compared to PC1

enteral nutrition with lipid-containing emulsions, especially the occurrence of stress cracking by intravenous infusion in three-way stop cocks made from PC could not be completely excluded.

A significant improvement in the administration of a lipid-containing emulsion was achieved with a special additive to PC. As shown in Fig. 22b, the improved behavior of the new polycarbonate PC2 proved in practice to be verifiable by fracture mechanical fatigue crack growth experiments. In the presence of the fat emulsion the more lipid-resistant PC2 shows a higher fatigue threshold value ΔK_{th} as well as an improvement by a factor of two in ΔK_{cf} in comparison to PC1 used so far.

3
Multiaxial Stress Loading

By definition, ESC is influenced by the level of the applied multiaxial stress. It is expected that below an assigned value of stress in a specific medium ESC will not occur. Fatigue crack growth experiments at various constant levels of ΔK can be applied to study the influence of the crack-tip loading on the ESC behavior in a systematic way. As an example, cast CT specimens of PMMA ($M_N = 4.6 \times 10^5$) were tested in air and IPA at different levels of ΔK (0.6, 0.7, 0.8, 0.9 MPa $\sqrt{m}$). The tests were performed as described earlier, with the application of the medium after a certain time of cyclic loading in air. From the plots of the crack length ratio vs. number of cycles, the following observations can be made (Fig. 23):

- At the lowest ΔK of 0.6 MPa $\sqrt{m}$, the injection of the medium causes the crack propagation rate to decrease rapidly (reduced slope following point A in Fig. 23a). After an incubation time of about 400 000 cycles (point B) the crack propagation speeds up suddenly and is followed by a stable crack propagation phase follwing point C with a slope representing a lower FCP rate as compared to propagation in air. Failure occurs at about 10^6 cycles.
- For $\Delta K = 0.7$ MPa $\sqrt{m}$ (Fig. 23b), after the injection of IPA crack propagation speeds up twice at point B and point D. In between, the increase in a/W is negligibly small. The stable crack propagation after point E is again slower than in air. Failure occurs at about 4×10^5 cycles.
- At higher levels of ΔK, 0.8 and 0.9 MPa $\sqrt{m}$ (Fig. 23c,d), the FCP characteristics change again. After the injection of IPA the crack propagation is quasistable, but at a lower propagation rate as compared to air. Despite stable crack extension at points B and D, sudden steps in the curve are visible, representing an area of unstable crack propagation over a very small number of cycles. The steps are more pronounced at ΔK 0.8 MPa $\sqrt{m}$.

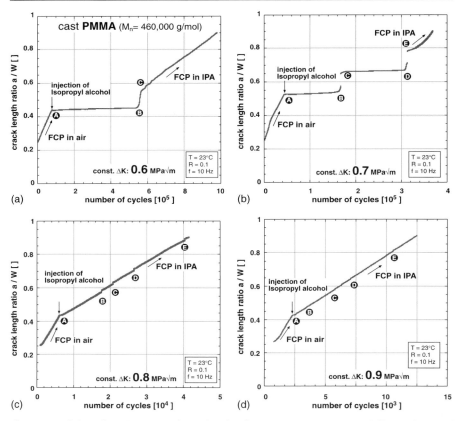

Fig. 23 Crack length ratio vs. number of cycles for cast PMMA in IPA at different levels of ΔK. ($a = 0.6$, $b = 0.7$, $c = 0.8$, $d = 0.9$)

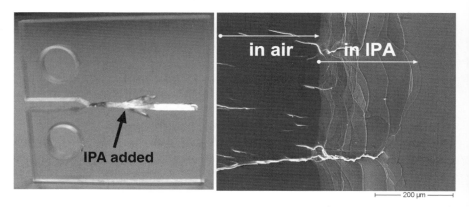

Fig. 24 Images of a cast PMMA sample after testing in IPA at $\Delta K = 0.6$ MPa $\sqrt{m}$ *Left*: Macroscopic view of crack branching along main crack front, *right*: SEM picture of fracture surface

In summary, the cast PMMA with high molecular weight shows better fatigue characteristics in IPA than in air. The reason for the apparent crack arrest is local "crack-tip branching", which can be located at the respective places along the crack where the stagnation in da/dN occurs. The branching of the crack does not contribute to the overall crack-tip opening and therefore cannot be detected directly. The discussed effect is documented by Fig. 24, left, which shows a CT sample tested at $\Delta K = 0.6$ MPa $\sqrt{m}$. The right-hand side of Fig. 24 depicts the branching by means of a scanning electron microscopy (SEM) picture.

4
Effect of Processing and Treatments

The method of preparation of a given polymer influences its mechanical properties, hence its ESCR. For amorphous polymers, orientations formed, e.g., in the injection molding process have a distinct influence upon fatigue under stress cracking conditions. Moreover, cross linking and aging also change the molecular state of the polymer, as will be shown in the following sections.

4.1
Orientation

When polymer melts are accelerated during the manufacturing processes such as injection molding and afterwards cooled down quickly, the stretched molecular chains cannot relax, which leads to orientations in the direction of melt flow. These oriented polymer chains act differently toward ESC influences as compared to nonoriented molecules. Orientation perpendicular to the crack propagation direction may result in a higher ESCR, as will be shown in the following discussion.

FCP tests of PS samples parallel and perpendicular to the injection molding direction were performed in comparison to material that was compression-molded from granules [76]. To minimize the effect of a plastic deformation possibly induced by precracking with a razor blade, the precrack was extended for a minimum of 2 mm by fatigue loading before onset of the experiment.

As shown in Fig. 25 for the performance in air, the FCP diagram of the specimens tested perpendicular to the injection direction is shifted significantly to higher ΔK values compared to the samples tested parallel to the injection direction. The nonoriented reference specimens, however, behave similarly to those with the parallel orientation. Obviously, it is easier for the fatigue crack to propagate in the direction of the oriented entanglement network. This can be explained by the fact that crazing accompanies crack propagation in PS.

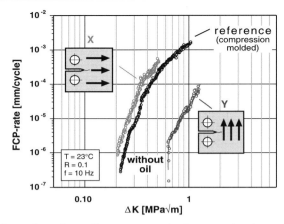

Fig. 25 Molecular orientation effects on FCP in lower-molecular-weight PS without medium (PS 148H)

Molecules that are already stretched in one direction lose the ability to fibrillate in the other direction. This increases the probability of chain scission, and the breakdown of the material drawn into the process zone occurs at lower ΔK values.

Figure 26 shows FCP curves for PS with a higher molecular weight tested perpendicular to the injection direction with and without oil. In oil, an additional embrittlement can be observed by the steep increase in the FCP curve, which makes it almost impossible to measure the dynamics of crack propagation over a broad range of crack propagation rates. Still, testing perpendicular

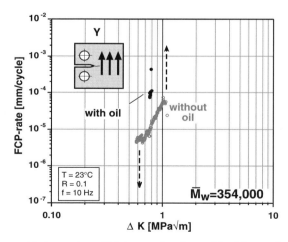

Fig. 26 Injection molded samples of higher-molecular-weight PS tested perpendicular to injection direction (PS 168N) with and without oil

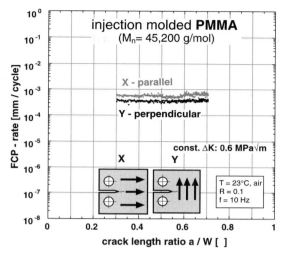

Fig. 27 Influence of orientation in PMMA (M_N = 45 200) induced by injection molding on crack propagation rate, depending on a/W

to the orientation results in higher ΔK values than for the parallel testing shown in Fig. 11.

Slightly different fatigue behavior due to orientation could also be observed in constant ΔK experiments performed on injection molded PMMA. The grade with M_N = 45 200 shows a lower crack propagation rate if the crack propagation is perpendicular to the injection molding direction (Fig. 27). Such a minor difference could also be found for these materials when the tests were performed with the stress cracking medium IPA. Because the molecular weight is expected to have a significant effect on the amount of orientation in a given shear stress field, the effect at a higher molecular weight must be investigated further.

4.2
Sterilization and Aging

Medical-grade ultra-high-molecular-weight polyethylene (UHMWPE) is a material of choice for orthopedic joints. The influence of sterilization, partly followed by accelerated aging as a factor of great practical importance, on its fatigue crack propagation behavior has been studied in cyclic tensile/compression tests [89]. Figure 28 shows the tensile crack propagation rates of unaged UHMWPE (GUR 4150HP) for different sterilization methods:

- No sterilization,
- Gas-plasma sterilization,
- Ethylene oxide sterilization,

- Gamma sterilization in air,
- Gamma sterilization in an inert environment.

Gas-plasma and ethylene oxide sterilization methods do not significantly reduce the fatigue resistance of the UHMWPE, while gamma radiation results in a substantial degradation, regardless of the sterilization environment. This ESCR decrease is due to the embrittlement resulting from the gamma radiation. The gamma sterilization in air treatment is associated with an initial oxidation and scission of the chains. In the absence of oxygen during ionizing radiation, cross-linking mechanisms are favored, although some degree of scission and oxidation still occurs.

Comparing Figs. 28 and 29, accelerated aging affects the fatigue resistance of all UHMWPE samples. Again, gas-plasma and ethylene oxide sterilization as well as the nonsterilized product show little differences after aging,

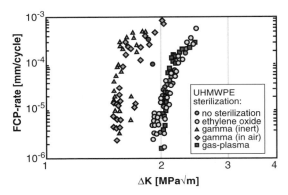

Fig. 28 FCP rates as function of stress intensity factor range for unaged UHMWPE (GUR 4150HP) following different sterilization methods

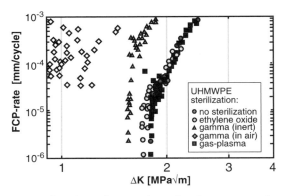

Fig. 29 FCP rates as a function of stress intensity factor range for aged UHMWPE (GUR 4150HP) following different sterilization methods

whereas the results for material gamma-irradiated in air and inert gas differ from each other. The most severe decrease in fatigue properties is due to the combination of gamma sterilization in air and accelerated aging since the oxidative degradation processes that were started by ionization continue during aging. The inert gamma treatment cannot induce oxidative species and therefore does not have a comparably strong effect on stress cracking.

5
Conclusion

Fatigue resistance of polymers in a stress cracking environment is a complex topic where molecular variables have a strong influence. FCP experiments are a fast and effective method for determining the resistance to FCP in stress cracking environments and hence to predict polymer lifetime. Other mechanical testing methods have also been cited as they are somewhat more common.

An increase in molecular weight leads to significant improvements in the fatigue life of both amorphous and crystalline polymers. This improvement is attributed primarily to an increase in craze stability and to an improvement in the fibril density, contributing to greater resistance to liquid transport. As the fraction of low-molecular-weight molecules cannot contribute to ESCR, it is found that for similar molecular weights narrow MWDs are clearly superior.

An increase in chain regularity, which contributes to a higher degree of crystallinity, leads to an improvement in ESCR due to the higher sliding resistance of the polymer chains through a crystal and through entanglements in amorphous regions. Short-chain branching is helpful for hindering disentanglement of molecules in stress cracking media; the best results can be found for relatively long chains, combined with a small-chain-length distribution. Rubber toughening as well as other polymer modifications (appropriate choice of stabilizers) are further means for improving ESC behavior.

Regarding fatigue performance, an important role is played by polymer processing. Molecular orientations and aging can affect polymers so that they show earlier failure in active media. On the other hand, appropriate orientations can be induced to enhance the ESCR.

Different strategies are available to increase the resistance of a polymer to ESC. These strategies include increasing the molecular weight, rubber modifications, and specific side-chain branching of the macromolecules. All these modifications aim at reducing craze formation, stabilizing the craze, and reducing the liquid diffusion in the craze.

Acknowledgements The author gratefully acknowledges the fruitful discussions with Henning Kausch and Falko Ramsteiner. In addition, the supply of experimental data on PMMA by Frank Fischer and the help in preparing the manuscript by Sandrine Müller,

Eva Bittmann, and Jan Sandler are acknowledged. Thanks are due to BASF AG and Röhm GmbH und Co KG for supplying the materials.

References

1. Wright DC (1996) Environmental stress cracking of plastics. Rapra Technology, Shawbury, ISBN: 1-85957-064-X
2. Klemchuk PP (1985) Enviromental degradation of polymers. In: Klemchuk PP (ed) Polymer Stabilization and Degradation. ACS Symposium Series 280, Washington, DC, pp 461–484. ISBN: 0-8412-0916-2
3. Wishnis C, Wright DC (1964) Adv Mater Process 12:93
4. Lustiger A (1986) Environmental stress cracking: the phenomenon and its utility. In: Brostow W, Corneliussen RD (ed) Failure of Plastics. Hanser Gardener, Munich, pp 305–329. ISBN: 1-56990-008-6
5. Goodwin AA, Whittaker AK, Jack KS, Hay JN, Forsythe J (2000) Absorption of low molecular weight penetrants by a thermoplastic polyimide. Polymer 41(19):7263–7271
6. Arnold JC (1995) The influence of liquid uptake on environmental stress cracking of glassy polymers. Mater Sci Eng A 197(1):119–124
7. Hansen CM, Just L (2001) Prediction of environmental stress cracking in plastics with Hansen solubility parameters. Ind Eng Chem Res 40(1):21–25
8. Hughes-Dillon K, Schroeder LW (1998) Stress cracking of polyurethanes by absorbed steroids. Polym Degradat Stabil 60(1):11–20
9. Arnold JC (1998) The effects of diffusion on environmental stress crack initiation in PMMA. J Mater Sci 33(21):5193–5204
10. Hansen CM (2002) On predicting environmental stress cracking in polymers. Polym Degradat Stabil 77(1):43–53
11. Arnold JC (1995) Craze initiation during the enviromental stress cracking of polymers. J Mater Sci 30(3):655–660
12. Turnbull A, Maxwell AS, Pillai S (2000) Comparative assessment of slow strain rate, 4-pt bend and constant load test methods for measuring environment stress cracking of polymers. Polym Test 19(2):117–129
13. Kramer EJ (1979) Environmental cracking of polymers. In: Andrews EH (ed) Developments in Polymer Fracture. Applied Science, London, pp 55–120. ISBN: 0-85334-819-7
14. Arnold JC (1994) The use of flexural tests in the study of enviromental stress cracking of polymers. Polym Eng Sci 34(8):665–670
15. Kefalas VA (1995) Solvent crazing as a stress-induced surface adsorption and bulk plasticization effect. J Appl Polym Sci 58(4):711–717
16. Marshall GP, Culver LE, Williams JG (1970) Craze growth in poly(methyl methacrylate): a fracture mechanics approach. Proc R Soc Lond A Math Phys Eng Sci 319(1537):165–187
17. Kramer EJ, Bubeck RA (1978) Growth kinetics of solvent crazes in glassy polymers. J Polym Sci Polym Phys 16(7):1195–1217
18. Kambour RP (1973) A review of crazing and fracture in thermoplastics. J Polym Sci Macromol Rev 7(1):1–154
19. Kambour RP, Gruner CL, Romagosa EE (1974) Bisphenol-A polycarbonate immersed in organic media: swelling and response to stress. Macromolecules 7(2):248–253

20. Bernier GA, Kambour RP (1968) The role of organic agents in the stress crazing and cracking of poly(2,6-dimethyl-1,4-phenylene oxide). Macromolecules 1(5):393–400
21. Kambour RP, Romagosa EE, Gruner CL (1972) Swelling, crazing, and cracking of an aromatic copolyether-sulfone in organic media. Macromolecules 5(4):335–340
22. Kambour RP, Gruner CL, Romagosa EE (1973) Solvent crazing of dry polystyrene and dry crazing of plasticized polystyrene. J Polym Sci Polym Phys 11(10):1879–1890
23. Kambour RP (1985) Crazing. In: Kroschwitz JI (ed) Encyclopedia of Polymer Science and Engineering. Wiley, New York, p 299. ISBN: 0-471-89540-7
24. Kambour RP, Gruner CL (1978) Effects of polar group incorporation on crazing of glassy polymers: styrene-acrylonitrile copolymer and a dicyano bisphenol polycarbonate. J Polym Sci Polym Phys 16(4):703–716
25. White SA, Weissman SR, Kambour RP (1982) Resistance of a polyetherimide to environmental stress crazing and cracking. J Appl Polym Sci 27(7):2675–2682
26. Hildebrand JH, Prausnitz TM, Scott RL (1970) Regular and Related Solutions. Krieger, New York. ISBN: 0-442-15665-0
27. Jacques CHM, Wyzgoski MG (1979) Prediction of environmental stress cracking of polycarbonate from solubility considerations. J Appl Polym Sci 23(4):1153–1166
28. Kawagoe M, Kitagawa M (1987) Craze initiation in poly(methyl methacrylate) exposed to n-alkanes. J Mater Sci 22(8):3000–3004
29. Kawagoe M, Morita M (1993) Fatigue failure of poly(methyl methacrylate) in alcohol environments. J Mater Sci 28(9):2347–2352
30. Kawagoe M, Kitagawa M (1981) Craze initiation in poly(methyl methacrylate) under biaxial stress. J Polym Sci Polym Phys 19(9):1423–1433
31. Lapique F, Meakin P, Feder J, Jøssang T (2002) Self-affine fractal scaling in fracture surfaces generated in ethylene and propylene polymers and copolymers. J Appl Polym Sci 86(4):973–983
32. Griffiths CL, Betteridge S (2000) The effect of environmental stress cracking on water tree growth. IEE Conference, 473 (Dielectric Materials, Measurements and Applications), pp 41–46
33. Lagarón JM, Capaccio G, Rose LJ, Kip BJ (2000) Craze morphology and molecular orientation in the slow crack growth failure of polyethylene. J Appl Polym Sci 77(2):283–296
34. Schellenberg J, Fienhold G (1998) Environmental stress cracking resistance of blends of high-density polyethylene with other polyethylenes. Polym Eng Sci 38(9):1413–1419
35. Dukes WA (1961) The Endurance of Polyethylene under constant tension while immersed in Igepal. British Plastics 34:123
36. Qian R, Lu X, Brown N (1993) The effect of concentration of an environmental stress cracking agent on slow crack growth in polyethylenes. Polymer 34(22):4727–4731
37. Ward AL, Lu X, Huang Y, Brown N (1991) The mechanism of slow crack growth in polyethylene by an environmental stress cracking agent. Polymer 32(12):2172–2178
38. Cho K, Lee MS, Park CE (1997) Environmental stress cracking of rubber-modified styrenic polymers in Freon vapour. Polymer 38(18):4641–4650
39. Cho K, Lee MS, Park CE (1998) The effect of Freon vapour on fracture behaviour of styrene-acrylonitrile copolymer – I. Craze initiation behaviour. Polymer 39(6–7):1357–1361
40. Maxwell AS, Turnbull A (2004) Influence of small fluctuating loads on environment stress cracking of polymers. Polym Test 23(4):419–422
41. Kawaguchi T, Nishimura H (2003) Environmental stress cracking (ESC) of plastics caused by non-ionic surfactants. Polym Eng Sci 43(2):419–430

42. Kawaguchi T, Nishimura H, Miwa F, Abe K, Kuriyama T, Narisawa I (1999) Environmental stress cracking of poly(acrylonitrile-butadiene-styrene). Polym Eng Sci 39(2):268–273
43. Wang HT, Pan BR, Du QG, Li YQ (2003) The strain in the test environmental stress cracking of plastics. Polym Test 22(2):125–128
44. Raman A, Farris RJ, Lesser AJ (2003) Effect of stress state and polymer morphology on environmental stress cracking in polycarbonate. J Appl Polym Sci 88(2):550–564
45. Al-Saidi LF, Mortensen K, Almdal K (2003) Environmental stress cracking resistance: behaviour of polycarbonate in different chemicals by determination of the time-dependence of stress at constant strains. Polym Degradat Stabil 82(3):451–461
46. Arnold JC, Taylor JE (1999) Improved thermodynamic approach for predicting the ESC behavior of polycarbonate in binary liquid mixtures. J Appl Polym Sci 71(13):2155–2161
47. Ishiyama C, Sakuma T, Shimojo M, Higo Y (2002) Effects of humidity on environmental stress cracking behavior in poly(methyl methacrylate). J Polym Sci Part B Polym Phys 40(1):1–9
48. Hough MC, Wright DC (1996) Two new test methods for assessing environmental stress cracking of amorphous thermoplastics. Polym Test 15(5):403–408
49. Martin DJ, Warren LAP, Gunatillake PA, McCarthy SJ, Meijs GF, Schindhelm K (2001) New methods for the assessment of in vitro and in vivo stress cracking in biomedical polyurethanes. Biomaterials 22(9):973–978
50. Arnold JC, Li J, Isaac DH (1996) The effects of pre-immersion in hostile environments on the ESC behaviour of urethane-acrylic polymers. J Mater Process Technol 56(1–4):126–135
51. Li J, Isaac DH, Arnold JC (1996) Environmental stress cracking in uniaxial tension of urethane methacrylate based resins. Mater Sci Eng A 214(1–2):68–77
52. Li J, Arnold JC, Isaac DH (1994) Environmental stress cracking behavior of urethane methacrylate based resins – I. Environmental crazing and cracking under bending conditions. J Mater Sci 29(12):3095–3101
53. Moskala EJ (1998) A fracture mechanics approach to environmental stress cracking in poly(ethyleneterephthalate). Polymer 39(3):675–680
54. Karger-Kocsis J (1991) Environmental stress corrosion behavior of polyamides and their composites with short glass fiber and glass swirl mat. Polym Bull 26(1):123–130
55. Bishop S, Isaac DH, Hinksman P, Morrissey P (2000) Environmental stress cracking of poly(vinyl chloride) in alkaline solutions. Polym Degradat Stabil 70(3):477–484
56. Hansen CM (2002) Environmental stress cracking of PTFE in kerosene. Polym Degradat Stabil 77(3):511–513
57. Maccone P, Brinati G, Arcella V (2000) Environmental stress cracking of poly(vinylidene fluoride) in sodium hydroxide: effect of chain regularity. Polym Eng Sci 40(3):761–767
58. Carter JT, Emmerson GT, Lo Faro C, McGrail PT, Moore DR (2003) The development of a low temperature cure modified epoxy resin system for aerospace composites. Compos Part A Appl Sci Manuf 34(1):83–91
59. Kawada H, Srivastava VK (2001) The effect of an acidic stress environment on the stress-intensity factor for GRP laminates. Compos Sci Technol 61(8):1109–1114
60. Bergen RL (1968) J SPE 24:667
61. Moskala EJ, Jones M (1994) Evaluating environmental stress cracking of medical plastics. Med Plast Biomater Mag 1(1):48–54
62. Arnold JC (1996) Environmental stress crack initiation in glassy polymer: review. Trends Polym Sci 4(12):403–408

63. Williams JG, Marshall GP (1975) Environmental crack and craze growth phenomena in polymers. Proc R Soc Lond Ser A 342(1628):55–57
64. Wyzgoski MG, Novak GE (1987) Stress cracking of nylon polymers in aqueous salt solutions. Part 3: Craze-growth kinetics. J Mater Sci 22(7):2615–2623
65. Wyzgoski MG, Novak GE (1987) Stress cracking of nylon polymers in aqueous salt solutions – Part II. Nylon-salt interactions. J Mater Sci 22(5):1715–1723
66. Hertzberg RW, Manson JA (1980) Fatigue of Engineering Plastics. Academic, New York. ISBN: 0-123-43550-1
67. Kausch HH (1987) Polymer Fracture. Springer, Berlin Heidelberg New York. ISBN: 0-387-13250-3
68. Suresh S, Pruitt L (1991) Fatigue crack growth in polymers and organic composites under cyclic compressive loads. In: Proceedings of the 8th International Conference on Deformation, Yield and Fracture Of Polymers, 32:1–4
69. Pruitt L, Hermann R, Suresh S (1992) Fatigue crack growth in polymers subjected to fully compressive cyclic loads. J Mater Sci 27(6):1608–1616
70. Pruitt L, Suresh S (1994) Cyclic stress fields ahead of tension fatigue cracks in amorphous polymers. Polymer 35(15):3221–3229
71. Pruitt L, Suresh S (1993) Cyclic stress fields for fatigue cracks in amorphous solids: experimental measurements and their implications. Philos Mag A 67(5):1219–1245
72. Saxena A, Hudak SJ (1979) Review and extension of compliance information for common crack growth specimens. Int J Fract 14(5):453–468
73. Sauer JA, Hara M (1990) Effect of molecular variables on crazing and fatigue. In: Kausch HH (ed) Advances in Polymer Science 91/92. Springer, Berlin Heidelberg New York, pp 69–118. ISBN: 3-540-51306-X
74. Castellani L, Rink M (2001) Fatigue crack growth of polymers. In: Moore DR, Pavan A, Williams JG (eds) Fracture Mechanics Testing Methods for Polymers. Adhesives and Composites. Elsevier, New York, pp 91–116. ISBN: 0-08-043689-7
75. Paris PC (1964) Proceedings of the 10th Sagamore Conference. Syracuse University Press, New York, p 107
76. Altstädt V (1997) Fatigue crack propagation in homopolymers and blends with high and low interphase strength. In: European Conference on Macromolecular Physics – Surfaces and Interfaces in Polymers and Composites, Lausanne, Switzerland
77. Altstädt V, Loth W, Schlarb A (1996) Comparison of fatigue test methods for research and development of polymers and polymer composites. In: Cardon AH, Fuduka H, Reifsnider KL (eds) Progress in Durability Analysis of Composite Systems. Balkema, Rotterdam, pp 75–80
78. Altstädt V, Keiter S, Renner M, Schlarb A (2000) Accelerated evaluation of environmental ageing monitored by fatigue crack growth experiments. In: Cardon AH, Fuduka H, Reifsnider KL, Verchery G (eds) Recent developments in durability analysis of composite systems. Balkema, Rotterdam, pp 195–209
79. Altstädt V, Keiter S, Renner M, Schlarb A (2004) Environmental stress cracking of polymers monitored by fatigue crack growth. Macromol Symp 214(1):31–46
80. Rudd RF (1963) Molecular-weight dependence of environmental stress cracking in polystyrene. J Polym Sci Part B Polym Lett 1(1):1–5
81. Wellinghoff S, Baer E (1975) Mechanism of crazing in polystyrene. J Macromol Sci Phys B 11(3):367–387
82. Rufke B (1992) Prüfung des Medienverhaltens. In: Schmiedel H (ed) Handbuch der Kunststoffprüfung. Carl Hanser, Munich, Vienna, pp 303–364. ISBN: 3-446-16336-0
83. Ramsteiner F (2005) Bewertung der Spannungsrissbeständigkeit. In: Grellmann W (ed) Prüfung von Verbundwerkstoffen; Kunststoffprüfung. Carl Hanser, Munich

84. Soares JBP, Abbott RF, Kim JD (2000) Environmental stress cracking resistance of polyethylene: the use of CRYSTAF and SEC to establish structure-property relationships. J Polym Sci Part B Polym Phys 38(10):1267–1275
85. Yeh JT, Jung-Horng Chen J-H, Huei-Song Hong H-S (1994) Environmental stress cracking behavior of short-chain branch polyehtylenes. In: Igepal Solution Under a Constant Load. J Appl Polym Sci 54:2171–2186
86. Maccone P, Brinati G, Arcella V (2000) Environmental stress cracking of poly(vinylidene fluoride) in sodium hydroxide: effect of chain regularity. Polym Eng Sci 40(3):761–767
87. Bubeck RA, Arends CB (1981) Environmental stress cracking in impact polystyrene. Polym Eng Sci 21(10):624–633
88. Pinter G, Lang RW (2001) Fracture mechanics characterisation of effects of stabilisers on creep crack growth in polyethylene pipes. Plast Rubber Compos 30(2):94–100
89. Baker DA, Hastings RS, Pruitt L (1999) Compression and tension fatigue resistance of medical grade ultra high molecular weight polyethylene: the effect of morphology, sterilization, aging and temperature. Polymer 41(2):795–808

Fracture of Glassy Polymers Within Sliding Contacts

Antoine Chateauminois[1] (✉) · Marie Christine Baietto-Dubourg[2]

[1]Laboratoire de Physico-Chimie des Polymères et des Milieux Dispersés, PPMD, CNRS UMR 7615, Ecole Supérieure de Physique et Chimie Industrielles (ESPCI), 10 rue Vauquelin, 75231 Paris Cedex 5, France
antoine.chateauminois@espci.fr

[2]Laboratoire de Mécanique des Contacts et des Solides, LAMCOS, CNRS UMR 5514, INSA de Lyon, 20 Avenue A. Einstein, Villeurbanne Cedex, France
Marie-Christine.baietto@insa-lyon.fr

1	Introduction	154
2	Overview of Experimental Approaches of Fracture Mechanisms Within Single Asperity Contacts	158
2.1	Micro/Nano Scratching Techniques	158
2.2	In Situ Contact Visualization Techniques	162
3	Contact Fatigue Behaviour of Epoxy Polymers	163
3.1	Contact Conditions Under Small Amplitude Micromotions	163
3.2	In Situ Analysis of the Cracking Behaviour	166
3.3	Crack Initiation: The Relevance of Bulk Fatigue Properties	171
3.4	Crack Propagation: Ductile-to-Brittle Transitions	178
4	The Role of Molecular Parameters in Contact Fracture Processes of Glassy Polymers	182
4.1	Toughened Epoxy Networks	182
4.2	Copolymers of Methylmethacrylate	185
5	Conclusion	189
	References	190

Abstract Cracking processes in brittle amorphous polymers within sliding contacts and their relationships with bulk fracture properties are reviewed. The focus is on the use of model single asperity contacts to mimic and characterize the failure modes which can be encountered at the microasperity level during the wear of macroscopic rough contacts between polymer surfaces and rigid counterfaces. Using the resources of in situ contact visualization, crack initiation and propagation mechanisms within epoxy substrates are detailed under contact fatigue conditions. With the prospect of understanding the fundamental mechanisms involved in particles detachment from brittle polymer surfaces, it is shown how cracks locations, orientations and depths can be predicted from a knowledge of the bulk toughness and fatigue properties of the polymer by means of a fracture mechanics analysis of the contact. In the last section, the sensitivity of contact fatigue processes to molecular parameters is addressed in the case of anti-plasticized epoxy networks and random copolymers of poly(methylmethacrylate).

Keywords Contact fatigue · Toughness · Fatigue properties · Wear

Abbreviations
- a Radius of the circular contact area
- K Tangential contact stiffness
- K_I Mode I stress intensity factor
- K_{II} Mode II stress intensity factor
- K_{IC} Mode I fracture toughness
- P Normal load
- Q Instantaneous value of the tangential load
- Q^* Maximum value of the tangential load
- δ Instantaneous value of the tangential relative displacement applied to the contact
- δ^* Maximum value of the cyclic relative displacement applied to the contact
- μ Coefficient of friction
- σ_H Hydrostatic stress
- τ_{y0}^{oct} Octahedral shear yield stress in the absence of hydrostatic stress
- $\sigma_{H.}$ Hydrostatic stress
- α Coefficient describing the sensitivity of the yield stress to hydrostatic pressure
- θ^* Calculated crack orientation with respect to the normal to the contact plane
- $\Delta\sigma_m^*$ Amplitude of the cyclic averaged effective tensile Stress perpendicular to the crack plane
- $\Delta\tau_m^*$ Amplitude of the average cyclic shear stress along the crack plane

1
Introduction

The topic of polymer wear and its relationships with mechanical properties has a long history that has been reviewed by several authors (see [1–6] for instance). From an historical point of view, wear processes within polymer materials have been essentially considered within the context of macroscopic contacts between rough surfaces sliding past each other. In such situations, the action of mechanical forces and/or chemical effects at discrete microcontacts between the surface asperities result in localized surface damage and the subsequent production of debris particles from one or both of the contacting surfaces. However, these failure processes are not monomechanistic and involve a sequence of often ill defined and interacting processes which are sensitive to several parameters such as contact loading conditions, surface roughness, material mechanical and physico-chemical properties, environment... Following early studies on the wear of metallic materials, phenomenological classifications based on some perceived judgments of the origins and consequences of particle detachment processes from polymer surfaces have been proposed in the literature [1, 2]. On the basis of microscopic observations of worn surfaces, "abrasion", "transfer wear", "chemical wear", "erosion", "fatigue wear", etc. have been pointed out as potential and often interacting wear modes. In order to rationalize the various descriptions of wear damage, Briscoe [3, 4] has proposed a generic scaling approach which emerges from the accepted value of the two-term non-interacting model of friction proposed by Tabor [7]. In this latter model, energy dissipation associated with friction, as a first approximation, is assumed to arise from two separate contributions: (1) an interfacial component (often denoted the "adhesive" component of friction), which is associated with the shearing of the adhesive junctions formed between contacting microasperities within macroscopic contacts, and (2) a bulk "ploughing" component which corresponds to the mechanical losses induced by the deformation of the softer material by the sliding asperities of the harder material.

If one considers that wear damage also results in energy dissipation, this approach can tentatively be applied to wear. Accordingly, wear processes can be classified as "cohesive" or "interfacial" depending on the length scales associated with particles detachment mechanisms (Fig. 1).

Adhesive, or *transfer wear* has long been recognized as a major source of wear in semi-crystalline polymers such as polytetrafluoroethylene (PTFE) and ultra high molecular weight polyethylene (UHMWPE), where friction is known to induce the transfer to the sliding counterface of highly oriented films a few tens of nanometres in thickness [8–13]. The shear work associated with the formation of such transfer films is dissipated within confined layers close to the sliding interface, where the strain rates and levels are far beyond those accessible by conventional bulk mechanical testing. In many cases, this naturally creates a significant increase in local temperature, which, in association with localized shear, can give rise to tribochemical degradation processes such as chain scission [14].

On the other hand, the general feature of *cohesive wear* is that frictional work is dissipated within a relatively thick surface zone either through the interaction of surface forces and the resultant traction stresses or simply via geometrical interlocking of interpenetrating contacts. Typically, these zones will be of a thickness which is of the order of a contact length. Within a macroscopic contact between rough surfaces, a small contact length of a few micrometres will be associated with the discrete asperity contacts and a longer length will be defined from the apparent (macroscopic) contact area. As opposed to interfacial modes, cohesive wear involves relatively mild viscoelastic or viscoplastic deformations which are to some extent accessible to bulk mechanical characterization, provided that the contact stress conditions are readily identified.

This crude distinction between adhesive and cohesive wear mechanisms is probably oversimplified in the sense that it neglects many aspects of the interactions between bulk deformation modes and interface rheology. It has, however, the merit of making a clear distinction between wear processes which can, to some extent, be related to known bulk failure properties and

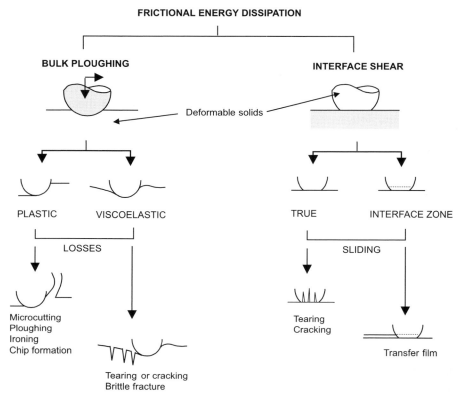

Fig. 1 Schematic description of "cohesive" and "interfacial" wear processes from the two terms non interacting model of friction (from [96]). Bulk ploughing involves the dissipation of the frictional work within a volume of the order of the cube of the contact radius. Interfacial shear corresponds to the dissipation of the frictional energy in much thinner regions and at greater energy densities. Cohesive wear processes (cracking, tearing, microcutting...) are governed by the cohesive strength of the polymer. Mechanisms such as transfer film formation correspond to interfacial wear and do not readily correlate with accessible bulk failure properties

more intense wear mechanisms which do not readily correlate with accessible bulk mechanical properties.

The relevance of bulk fracture properties has therefore been considered essentially within the context of cohesive wear modes such as abrasive and fatigue wear. During abrasive wear, the initial stage is considered to be the process of contact and scratch between the polymer surface and a sharp asperity. The accumulation of the associated microscopic failure events eventually generates wear particles and gives rise to weight loss. Early approaches initiated by Ratner and co-workers [15] and Lancaster [16] attempted to correlate the abrasive wear rate with some estimate of the work to failure of the same polymer. Using several semi-crystalline and glassy amorphous polymers with tensile works to rupture ranging over more than two orders of magnitude, they indeed showed the existence of a roughly linear relationship between the measured abrasive wear rates and the reciprocal of the work to failure. Owing to the lack of information regarding the nature of the microscopic deformation and failure processes, several doubts surround, however, the value of this correlation when a more detailed analysis of the relationships between abrasive wear and bulk mechanical properties is considered. It is especially unclear whether fatigue properties rather than the fracture energy involved in a single cracking event should be considered. The occurrence of fatigue processes induced by the repeated sliding of rigid asperities on elastically deforming polymer surfaces has indeed often been pointed out as a potential source of wear damage in many brittle polymers. Accordingly, macroscopic fatigue wear models have been derived [2, 17–19] which attempt to correlate the wear rates with an estimate of the number of cycles to detach a particle from the polymer surface. The latter is usually assessed from a knowledge of bulk fatigue properties and from some estimate of the distribution of the localized tensile stresses which are generated at the microasperity level. Alternatively, models of fatigue wear approach the problem from the standpoint of fracture mechanics [2, 17]. The wear rate is assumed to be inversely proportional to the number of cycles to failure and this can be derived by integrating the Paris equation for fatigue crack growth. Although such models can account for the effects of parameters such as surface roughness or environment in stress corrosion cracking conditions, the lack of information regarding crack orientation and propagation rates at the microscopic level precludes any more quantitative analysis of fatigue wear processes.

In addition to the difficulties associated with the identification of the deformation and failure processes, the analysis of wear processes within macroscopic contacts between rough surfaces can also be complicated by the behaviour of the wear debris trapped at the sliding interfaces. Once detached, the debris particles are invariably subjected to a variety of processes such as comminution, chemical reactions, aggregation and compaction and ultimately form what is often called a "third body" [20, 21]. It is now widely acknowledged that the third body rheology at the sliding interface can significantly affect the mechanical stressing of the rubbing surfaces and modify the associated particle detachment mechanisms. In poly(methylmethacrylate)/steel contacts, it has, for example, been shown that specific contact zone kinematics conditions can result in the accumulation of polymeric debris particles within distinct and highly coherent third body agglomerates [22–24]. Such situations are associated with a strong decrease in the wear damage as a substantial part of the frictional energy is dissipated through the shearing of the third body which acts as a solid lubricant.

These observations underline the strongly intricate interactions between particle detachment, third body rheology and evolving contact conditions which are inherent to wear processes in many macroscopic contacts.

In order to overcome the above-mentioned inherent limitations of macroscopic, multi-asperity contacts for the fundamental investigation of wear processes, model experiments that attempt to simulate the damage induced by a single asperity contact are often considered. Although the wear rate itself is not monitored, such experiments provide the opportunity to get a more detailed insight into the deformation and fracture mechanisms involved in asperity engagements [25]. The recent emergence of hardness and nano-scratching probes has provided the opportunity to investigate these processes at length scales which are comparable with those involved in typical sliding contacts. Alternatively, single asperity contacts at larger scales offer the possibility of investigating the dynamics of contact failure using the resources of in situ visualization. Both approaches will be reviewed in the subsequent sections, with an emphasis on the use of single asperity contact to investigate contact fatigue processes in relation to fatigue and toughness properties.

Within this context, the analysis will be focused on brittle amorphous polymers such as epoxies and acrylates, whose tribological damage in sliding contacts is known to be dominated by cracking micromechanisms. Such materials are relevant to many tribological applications. Fibre-reinforced composites with thermoset matrices are, for example, used as seal gear and dry bearing materials [26, 27]. Similarly, contact deformation and fracture processes of acrylate-based coatings can be a concern regarding wear and scratch resistance properties of optical coatings and varnishes [28–30]. Within the context of this review, these materials will, however, mostly be considered as model polymer systems which allow some insight into the fundamental mechanisms of crack nucleation and growth within contacts. Such an approach would be much more complicated using more ductile engineering polymers for tribological applications, such as polyamides or polyoxymethylene, where the in situ analysis of cracking processes is limited by transfer film formation and extensive plastic flow.

2
Overview of Experimental Approaches of Fracture Mechanisms Within Single Asperity Contacts

2.1
Micro/Nano Scratching Techniques

Since the early work from Schallamach [31] on the wear processes of rubbers, indentation and scratching experiments have been largely used to mimic and characterize the deformation and failure modes which occur as a consequence of specific interactions between asperities touching each other in discrete areas of the contact between rough polymeric solids. Although it does not model the dynamic asperity interactions encountered in practical tribological contacts, any form of scratch hardness can be viewed as a form of controlled abrasive wear. Extrapolation of scratch tests to wear situations must, however, be done with some caution as the geometry of the scratch indenter is likely to be different (usually very much more severe) from those of the asperities that make up the topography of even comparatively rough surfaces. Moreover, scratch tests are unable to reproduce fatigue failure events, which have long been recognized as a potential source of particle detachment in many practical wear situations involving polymers.

From a more fundamental point of view, the selection of different indenter geometries and loading conditions offer the possibility of exploring the viscoelastic/viscoplastic response and brittle failure mechanisms over a wide range of strain and strain rates. The relationship between imposed contact strain and indenter geometry has been quite well established for normal indentation. In the case of a conical or pyramidal indenter, the mean contact strain is usually considered to depend on the contact slope, θ (Fig. 2a). For metals, Tabor [32] has established that the mean strain is about $0.2 \tan\theta$, i.e. independent of the indentation depth. A similar relationship seems to hold for polymers although there is some indication that the proportionality could be lower than 0.2 for viscoelastic materials [33, 34]. In the case of a sphere, an

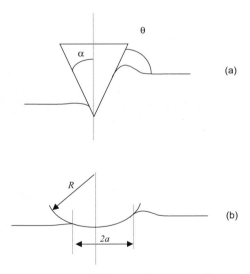

Fig. 2 Typical contact geometries used for scratch testing. The average strain is proportional **a** to $\tan\theta$ in the case of conical or pyramidal indentors and **b** to a/R in the case of spherical indentors

indentation depth-dependent strain can be defined as the ratio of the contact radius to the sphere radius (Fig. 2b). Similarly, an average strain rate can be defined by dividing the sliding speed by some characteristic length, usually the width of the permanent groove left on the polymer surface. Typical mean strain values during scratch tests lie in the range 5–10% and the strain rate can be varied from 10^{-4} s^{-1} to 10^2 s^{-1} in most of the commercially available scratching equipments.

Topographical characterizations using SEM, AFM or laser reflectivity are usually associated with tangential force measurements in order to assess the various regimes of deformation. The observed damage evolves through a range of severity as the contact strain is increased: viscoelastic smoothing or "ironing", plastic or viscoplastic grooving, extensive plastic flow and tearing, pronounced fracture or tearing and finally cutting or chip formation can be identified using glassy polymers such as poly(methylmethacrylate) or polycarbonate.

These approaches have been especially popularized for a variety of amorphous and glassy and semi-crystalline polymers by Briscoe and coworkers [25, 34–37], who put together in the form of "deformation maps" the different deformation regimes. The indenter angle and the penetration depth (or normal load) are usually considered the key parameters. Figure 3 provides an example of such a scratching map for polycarbonate using a conical indenter. It can be seen that, taking an arbitrary constant load, the nature of the contact damage can change from viscoplastic deformation ("ductile ploughing") to brittle crack propagation when the indenter angle is decreased, i.e. when the contact strain is increased.

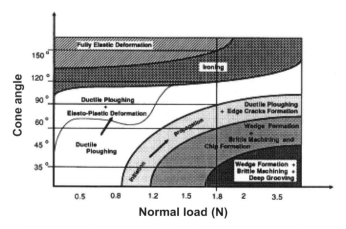

Fig. 3 Scratching map for poly(carbonate). The diagram shows results from scratching experiments performed at room temperature for a range of cone angles and normal loads and at a scratching velocity of 2.6 μm s^{-1} (from [25])

For a variety of glassy polymers, it is usually observed that cracks are induced at the leading edge of the contact under the action of predominantly tensile stresses, whose magnitude is assumed to be proportional to the frictional forces (Fig. 4). Accordingly, several authors have attempted to correlate this ductile–brittle transition to the tensile strength of the polymer material under investigation. Although some correlations emerge when parameters such as temperature, strain rate or indentation depth are considered [38–41], this analysis remains obscured by two factors. The first is connected to the difficulties in assessing the magnitude of the tensile stress field at the leading edge of the contact under largely viscoelastic/plastic conditions. A first order estimate can be obtained from well established elastic contact mechanics approaches (such as Goodman's theory [42], see below), but more precise calculations require the complexities of finite elements simulations which take into account strain hardening effects and viscoelastic recovery at the rear face of the indenter [43, 44].

Some difficulties also arise for the interpretation of scratch tests carried out at progressively increasing normal load or indentation depth. Figure 3 indicates, for example, that a transition from ductile deformation to brittle cracking can occur when increasing the normal load whilst the contact strain is nominally fixed by the conical indenter angle. This is indeed observed in many polymer systems and the notion of a critical load at the ductile–brittle transition is largely used to characterize the scratch response. This depth

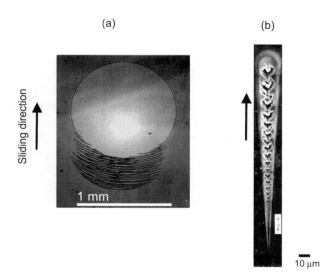

Fig. 4 Brittle failure modes of polystyrene within contacts. **a** Poly(styrene) film on a poly(methylmethacrylate) substrate. The regular crack pattern is induced by the sliding of a glass sphere under elastic contact conditions. **b** Poly(styrene) under viscoplastic scratching by a cone indenter (from [40])

dependence is usually interpreted by considering the enhancement of surface tensile stresses which may arise from some increase in the frictional forces at increased indentation depth, but some recent analysis by Bertrand-Lambotte et al. [45, 46] indicates that some size effects could also be involved in the transition. According to an argument initially introduced by Puttick et al. [47, 48] for the analysis of the ductile–brittle transition during ball indentation of poly(methylmethacrylate), these authors suggested that the occurrence of brittle scratches can be interpreted using two different criteria. The first one is a fracture energy criterion which states that a crack of width c, subjected to a tensile stress, σ_t, will grow if $\sigma_t^2 c/E \geq G_c$, where E and G_c are respectively the elastic modulus and the fracture energy. The second criterion is a size criterion which assumes that within a given volume, a^3, containing a crack of length, c, brittle failure will occur if the energy for crack propagation is less than that required for plastic deformation, i.e. $2G_c/\sigma_y\varepsilon_y \leq a$, where σ_y and ε_y are respectively the yield stress and plastic strain. According to Bertrand-Lambotte et al., the occurrence of brittle scratch, attributed by most authors to a critical load can be due to the fulfillment of either the fracture energy criterion *or* the size criterion. An estimate of the latter criterion for polymeric coatings reveals that the critical size at the ductile/brittle transition is of the order of a few micrometres, which corresponds to the characteristic scratch width at the onset of brittle propagation using pyramidal indenters.

2.2
In Situ Contact Visualization Techniques

When using transparent indenters and/or polymer materials, in situ contact visualization proved to be an efficient technique to investigate the dynamics of deformation and failure processes within sliding contacts. Such approaches have been recently developed within the context of scratch tests in order to investigate the elastic recovery of scratch grooves left on the surface of poly(methylmethacrylate) [49, 50] at the micrometre scale. They proved to be especially useful for analysing the transition from viscoplastic scratching to elastic sliding as a function of temperature, strain and strain rate.

The resources of in situ contact visualization have also recently been used to investigate crack initiation and propagation within brittle amorphous polymers under contact fatigue conditions. A detailed insight into the development of surface fatigue cracks was provided by means of specific tribological tests which rely upon the use of small-amplitude oscillating micromotions, often referred to as "fretting". In such a technique, a millimetre-sized macroscopic contact is typically formed between a polymer flat and a smooth spherical counterface under a constant applied normal force. This contact is submitted to lateral relative microdisplacements under cyclic conditions. As compared to other more conventional reciprocating sliding tests, the specificity here is that the magnitude of the imposed relative displacement (between about 1 and 100 μm) is much smaller than the size of the contact. In other words, this means that the contact area is nearly stationary with respect to the polymer surface, which considerably simplifies the analysis of crack development as compared to large amplitude sliding tests where multiple crack interactions have to be taken into account [51, 52].

Such an in situ visualization of fretting contacts demonstrated that it was possible to induce, at a length scale of millimetric, the development of a crack network without the unwanted complexities which could arise from the simultaneous development of an intercalating third body layer [53]. From the point of view of polymer fracture processes, this means that such fretting contacts may be viewed as some kind of model single-asperity contacts, which simulate, at an observable scale, contact fatigue processes. The use of this methodology to investigate some of the aspects of the structure/contact fracture properties relationships is addressed in the following sections.

3
Contact Fatigue Behaviour of Epoxy Polymers

3.1
Contact Conditions Under Small Amplitude Micromotions

The establishment of the interrelationships between polymer fracture properties and contact fatigue behaviour requires some knowledge of the contact stress field. The specificity of fretting loading is that, depending on the applied contact load and imposed relative displacement, two different contact conditions can be achieved within the interface [54, 55] (Fig. 5):

(1) The so-called partial slip condition, where the contact area is divided into a central stuck zone and a surrounding annulus where some microslip is occurring during the course of the loading cycle. When the tangential load is plotted as a function of the relative displacement in a Lissajous representation, elliptical loops are obtained.

(2) A gross slip condition, where sliding is induced within the whole contact interface after a preliminary partial slip stage. This condition is associated to trapezoidal tangential load/displacement loops. The plateau value of the tangential load provides a measurement of the coefficient of friction, $\mu = Q^*/P$, where Q^* and P are the plateau value of the tangential load and the imposed normal load respectively.

The occurrence of either partial slip or gross slip condition is dependent on the material mechanical properties, the magnitude of the coefficient of friction and the contact loading parameters (normal load, imposed displacement). When dealing with non-adhesive elastic materials, the effects of these

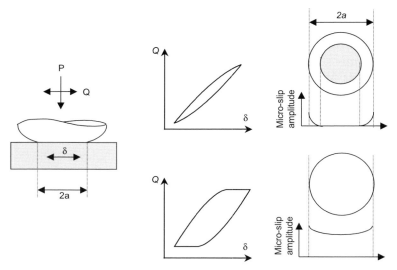

Fig. 5 Schematic description of the contact conditions encountered under small amplitude cyclic lateral micro-motions (fretting). δ is the applied lateral displacement, Q is the lateral force and P is the applied constant normal load. The elliptic and trapezoidal $Q(\delta)$ loops correspond to partial slip and gross slip condition respectively

factors can be rationalized by means of some conventional contact mechanics approach such as Mindlin's analysis [54]. This approach in particular provides the value of the critical displacement amplitude, δ_t, at the transition from partial slip to gross slip conditions:

$$\delta_t = \frac{K_1 \mu P}{a} \qquad (1)$$

with

$$K_1 = \frac{3}{16}\left(\frac{2-\nu_1}{G_1} + \frac{2-\nu_2}{G_2}\right) \qquad (2)$$

where the subscripts 1 and 2 refers to the contacting bodies and G and ν are respectively the shear modulus and the Poisson's coefficient.

With polymers, complications may potentially arise due to the material viscoelastic response. For glassy amorphous polymers tested far below their glass transition temperature, such viscoelastic effects were not found, however, to induce a significant departure from this theoretical prediction of the boundary between partial slip and gross slip conditions [56].

During the course of a cyclic test, some changes in the coefficient of friction may arise as a consequence, for example, of the development of physicochemical interactions between the contacting surfaces. For glass in contact with glassy polymers such as poly(methylmethacrylate) or epoxies, an in-

Fracture of Glassy Polymers Within Sliding Contacts

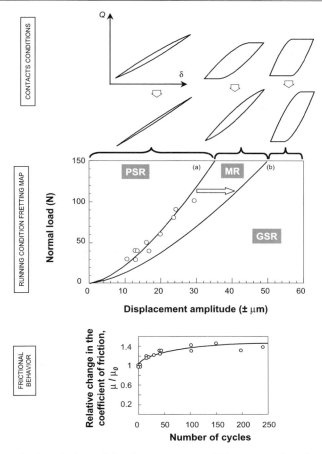

Fig. 6 Schematic description of fretting contact conditions as a function of displacement amplitude, normal load and number of loading cycles for a glass/epoxy contact (from [56]). *PSR* Partial slip regime, *MR* mixed regime, *GSR* gross slip regime. *Solid lines* in the running conditions fretting maps correspond to the theoretical boundary between partial slip and gross slip conditions, as calculated from Eq. 2 and using the initial (a) and steady-state (b) values of the coefficient of friction. *Open symbols* correspond to the experimental values of the initial boundary between partial slip and gross slip conditions. The relative increase in the measured coefficient of friction, μ, as a function of the number of fretting cycles is reported in the *lower figure* (μ_0 is the initial value of the coefficient of friction). The resulting shift of the partial slip/gross slip boundary defines the mixed regime which is characterized by a change from gross slip to partial slip condition during the course of the fretting test

crease in the coefficient of friction is often observed during some preliminary running-in period, even in the absence of any observable contact damage. One of the effects of this time-dependent frictional behaviour is to induce a progressive transition from gross slip to partial slip conditions in some intermediate displacement regime located close to the initial boundary between the partial slip and gross slip conditions.

As a conclusion, contact conditions under fretting loading can be classified into three different regimes: a partial slip regime, a gross slip regime and a mixed regime which is characterized by the progressive transition from gross slip to partial slip conditions during the course of the cyclic test. The occurrence of these various regimes can conveniently be synthesized as a function of the normal load and the imposed tangential displacement in the so-called running fretting condition maps [57, 58] (Fig. 6), where the boundaries between the various domains can be delimited by means of Eq. 1. As detailed below, the identification of these various regimes is of primary importance to set up the boundary conditions for the contact mechanics modelling of the stress field induced by the cyclic loading. From the point of view of contact fatigue processes, the fretting maps also emphasize the fact that cracking processes are potentially induced under a non-constant amplitude fatigue loading, as opposed to current practice for bulk fatigue testing.

3.2
In Situ Analysis of the Cracking Behaviour

In this section, contact fatigue cracking mechanisms will be reviewed in the case of an epoxy network obtained by fully cross-linking a stoichiometric mixture of diglycidyl ether of bisphenol (DGEBA) and isophoron diamine (IPD). It was, however, observed that the nature of the various cracking processes remains fundamentally unchanged when various glassy epoxy networks differing in their chemical structure are considered; the main differences lies in crack initiation times and crack growth rates. The results described here for the DGEBA/IPD network can thus be considered a picture of generic contact fatigue processes within brittle epoxy networks tested in their glassy state.

A typical example of the development of a crack network within the gross slip regime is shown in Fig. 7. In addition to in situ visualization through the glass cap, the crack growth rates were also monitored by means of measurements of the contact lateral stiffness, K. As indicated in the insert in Fig. 7, this latter parameter corresponds to the slope of the $Q(\delta)$ relationship during the incipient stages of the tangential loading, i.e. in the limit of small relative displacements. Under such conditions, one can neglect the effects of partial slip and describe the tangential loading as the drag of a circular region of the sample surface by the slider. In the case of an elastic, uncracked body, the associated lateral stiffness is known to be proportional to the contact radius and the elastic modulus of the materials [54]. When a crack propagates in the contact zone, a significant drop in the measured contact stiffness can therefore be expected by virtue of

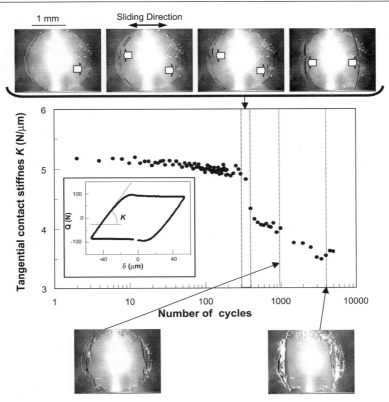

Fig. 7 Development of fatigue cracks in an epoxy/glass contact under gross slip condition (1 Hz, displacement amplitude: ±60 μm) (from [97]). *White arrows* indicate the occurrence of crack initiation and propagation at the edge of the contact under the action of tensile stresses. The lateral contact stiffness, K, is essentially a measurement of the elastic response of the epoxy substrate within the contact zone. Brittle crack propagation is associated to a drop in stiffness due to the additional accommodation of the imposed displacement provided by crack opening mechanisms

the reduction in the stiffness of the cracked body which results from crack opening mechanisms. As detailed below, this effect is particularly marked when the crack depth becomes of the order of magnitude of the contact radius.

Three successive stages can schematically be distinguished during the course of a contact fatigue test:

(1) A crack initiation stage which corresponds to the nucleation of two separate surface cracks close the edge of the contact and at two approximately symmetrical locations along the sliding direction. Within the resolution of the optical device (about 10 μm), crack nucleation was detected after about 300 loading cycles. During the next 100 fretting cycles, the cracks propagate in

progressive, fatigue-like manner along a direction roughly perpendicular to the sliding direction.

(2) When some critical crack length (between 200 μm and 400 μm, i.e. about 15% of the contact diameter) is reached, a brittle propagation stage is observed which is associated with a sudden and drastic drop in the lateral stiffness, K. The measured crack width in the plane of the contact is then of the order of magnitude of the contact diameter. Post-mortem microscope observation of specimen cross sections in the contact zone (Fig. 8) indicates that the depth of the cracks is of the order of magnitude of the contact radius (i.e. about 900 μm). The two deep cracks induced at the edge of the contact may thus be viewed as some kind of "half-penny" cracks whose radii are approximately equal to the radius of the contact. In the subsequent part of this paper, the two deep cracks will be referred to as "primary cracks".

(3) After the brittle propagation stage, "secondary cracks" are subsequently induced in the vicinity of the two primary cracks, but they result in a more limited and progressive decrease in the contact stiffness. Observation of contact cross-section shows that the depth of these cracks is strongly reduced compared to those of the primary cracks. After the propagation of the primary cracks, stresses are relaxed in their vicinity as a consequence of crack opening mechanisms, which can account for the limited propagation of the secondary cracks.

A simple contact mechanics analysis shows that the observed crack network at the vicinity of the contact edge is essentially induced under the action of alternate tensile and compressive stresses oriented along the sliding direction. According to acknowledged analytical expressions for the stress field induced by a sliding sphere on an elastic substrate [42], it appears that the complex multiaxial contact stress field reduces to a predominantly tensile/compressive loading in the region where crack nucleation is experimentally detected. An estimate of the maximum value of the tensile stress, σ_{xx}^{max}, experienced at the trailing edge of the contact is provided by the follow-

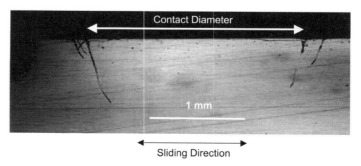

Fig. 8 Post mortem optical observation of a contact cross-section showing crack depth and orientation within an epoxy polymer after contact fatigue

ing formula:

$$\sigma_{xx}^{max} = \frac{3P}{2\pi a^2}\left[\frac{1-2\nu}{3} + \frac{4+\nu}{8}\pi\mu\right]. \quad (3)$$

Where P is the applied normal load, a is the contact radius, ν is the Poisson's ratio and μ is the coefficient of friction. For the above-described experiments,

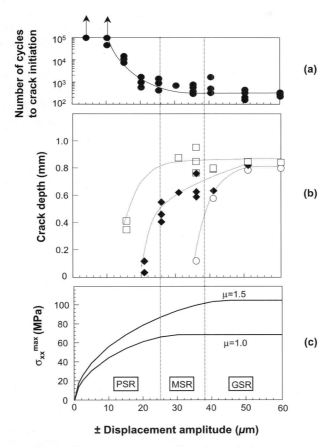

Fig. 9 Changes in the crack initiation times and crack depths in an epoxy resin as a function of the amplitude of the imposed cyclic displacement. **a** Number of cycles to the initiation of the primary cracks at the edge of the contact zone. **b** Measured depths of the primary cracks at various number of cycles and displacement amplitudes. *Circles* 10^3 cycles, *solid diamonds* 5×10^3 cycles, *squares* 5×10^4 cycles. **c** Calculated values of the maximum tensile stress at the edge of the contact using Hamilton (gross slip condition) or Mindlin—Cattaneo (partial slip condition) theories. The two curves correspond to calculations using the initial ($\mu = 1.0$) and the steady-state ($\mu = 1.5$) values of the coefficient of friction. *PSR* Partial slip regime, *MR* mixed regime, *GSR* gross slip regime

the calculated values of σ_{xx}^{max}, are in the range 50–100 MPa depending on the value of the coefficient of friction, which is consistent with the development of fatigue/brittle failure processes.

The characteristic features of the crack network remain essentially unchanged within the partial slip and mixed regimes: primary and secondary cracks are successively induced at the edge of the contact but, as opposed to the gross slip regime, both the number of cycles to crack initiation and the crack growth rates appear to be strongly sensitive to the magnitude of the applied relative displacement (Fig. 9). The more the imposed displacement is reduced, the more crack initiation processes are delayed. For approximately ±10 µm, a contact endurance limit can be identified below which no crack is nucleated after 10^5 cycles. In addition to these increased crack nucleation times, the measured crack depths (i.e. perpendicular to the contact plane, Fig. 9) and crack lengths (i.e. in the contact plane, Fig. 10) are strongly reduced at low displacement amplitudes (Fig. 10). Within the partial slip regime, it is also no longer possible to detect any brittle failure event as it is the case in the mixed and gross slip regimes (Fig. 10).

Under partial slip conditions, the estimate of the stress conditions at the edge of the contacts is complicated due to the unknown frictional behaviour within the partial slip annulus. If Coulomb's friction law is assumed to apply locally within this area, some contact mechanics calculation can, however, be

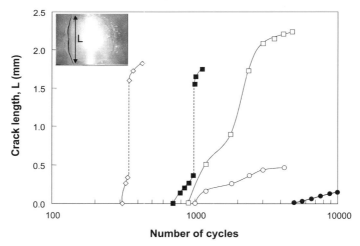

Fig. 10 Measured crack lengths, L, in an epoxy polymer as a function of the number of cycles and the imposed lateral displacement. The crack length was measured in the contact plane from in situ observations. Amplitude of the relative displacement: *Black circles* ±15 µm, *white circles* ±20 µm, *white squares* ±25 µm, *black squares* ±30 µm, *white diamonds* ±60 µm. *Dotted lines* indicate the occurrence of brittle failure events during the course of the contact fatigue experiments

carried out using conventional Mindlin–Cattaneo [54, 59] approaches. Such calculations show that the tensile/compressive nature of the stress field at the edge of the contact remains essentially unchanged as the amplitude of the displacement is decreased within the partial slip regime. As shown in Fig. 9, a significant decrease in the magnitude of the maximum value of the tensile stress is also predicted when the displacement amplitude is reduced, which can account for the observed reduction in the extent of the contact cracking processes.

3.3
Crack Initiation: The Relevance of Bulk Fatigue Properties

The fact that contact cracks are induced under a predominantly uniaxial tensile loading makes it realistic to attempt to establish some relationships with bulk fatigue behaviour under tensile conditions. Fatigue data for neat epoxy resins are scarce in the literature, but tend to indicate that, for unnotched specimens, most of the fatigue life is often consumed during crack initiation processes which take place close to the specimen surface [60, 61]. Such an observation is consistent with the above-described contact fatigue behaviour, especially in the mixed and gross slip regimes, where it was observed that crack initiation was rapidly followed by some brittle failure events.

When crack initiation is considered, some questions arise regarding the potential dependence of such processes on the stressed polymer volume, which largely differs between bulk fatigue testing and contact fatigue (in the latter case, the stressed volume is of the order of magnitude of the cube of the contact radius, i.e. a few mm^3). Such volume effects could be very significant in situations where crack initiation involves randomly distributed pre-existing flaws such as voids or inclusions. The mechanisms for crack initiation in amorphous polymers such as epoxies are still largely unclear, but they do not seem to involve any detectable pre-existing flaws, voids or inclusions [60, 61]. Observations of fracture surfaces show that fatigue cracks are usually nucleated on the specimen surface after a period which is close to the time to fracture. Nagasawa et al. [60] have also made the interesting observation that if a specimen is allowed to rest overnight when the fatigue test has proceeded by about 80% of the expected number of cycles to failure, the fatigue life of the specimen is subsequently increased by the corresponding number of fatigue cycles. Moreover, if such resting periods are alternated with successive fatigue cycling sequences, there is some indication that the fatigue life is considerably increased. These observations were tentatively interpreted by Nagasawa et al as showing that crack initiation is associated with some kind of recoverable strain accumulation close to the specimen surface. In an early work, Rabinowitz and co-workers [62] also showed that crack nucleation during fatigue testing of amorphous polymers such as polycarbonate is preceded by some strain-activated changes in the cyclic stress–strain response (cyclic strain softening) which were assumed to involve the nucleation and growth of microscopic defects within the glassy polymer. All these observations tend to support the idea that the relevant length scale for crack nucleation processes lies in the microscopic range, probably even at some molecular level. This means that no significant dependence of crack initiation on the stressed volume is likely to occur and that the use of bulk fatigue data should be relevant in the context of contact crack initiation.

More importantly, the hypothesis of crack nucleation processes involving some strain or molecular defects accumulation within the polymer addresses the question of the elastic or plastic nature of the deformation at the crack initiation location. By virtue of the strong heterogeneities of the contact loading, some localized plasticity may occur even if the contact remains, as a whole, loaded elastically. The occurrence of plastic deformation in the contact zone can be analysed using some multiaxial plasticity criteria. When considering the use of such criteria for polymers, some care must be taken, regarding the well known effects of hydrostatic pressure on the yield behaviour of glassy polymers. Owing to the confinement of the polymer material within the contact zone, such effects are likely to be significant and some modified form of conventional yield criteria have to be considered. In the case of epoxy materials subjected to a multiaxial stress field under constrained states of stress Lesser et al. [63, 64] demonstrated the validity of a modified Von Mises criterion which can be expressed as a function of the octahedral shear stress, τ_y^{oct}:

$$\tau_y^{oct} = \tau_{y0}^{oct} - \alpha \sigma_H \tag{4}$$

where σ_H is the hydrostatic stress and τ_{y0}^{oct} is the octahedral shear yield stress in the absence of σ_H. α is a coefficient that quantifies the hydrostatic pressure dependence of the yield limit. The octahedral shear stress, τ^{oct} and σ_H can be expressed as a function of the principal stresses as follows:

$$\tau^{oct} = \frac{1}{3} \left[(\sigma_1 - \sigma_2)^2 + (\sigma_1 - \sigma_3)^2 + (\sigma_2 - \sigma_3)^2 \right]^{1/2} \tag{5}$$

$$\sigma_H = -\frac{1}{3}(\sigma_1 + \sigma_2 + \sigma_3) \tag{6}$$

Under gross slip conditions, the distribution of octahedral shear stress and pressure within the contact can be calculated from the explicit theoretical expressions of the contact stresses beneath a rigid sliding sphere which were derived by Hamilton [42]. As detailed in [63, 65], the two parameters, τ_{octo} and α can be obtained from an appropriate combination of bulk mechanical tests such as uniaxial and plane strain compression tests. For practical reasons, the equivalent strain rates used during the bulk mechanical testing are usually lower than the strain rates achieved within the epoxy surface layer during contact fatigue (of the order of 10^{-2} s^{-1} at 1 Hz). Accordingly, the values of the octahedral shear stress at the onset of yield are probably slightly underestimated, although the strain rate dependence in the glassy state is not likely to be very important well below the glass transition temperature.

As an example, a calculated octahedral shear stress profile is reported in Fig. 11 that corresponds to the surface of the epoxy specimen, where the maximum value of the octahedral shear stress is likely to occur. This figure indicates that the contact remains elastic for the contact conditions under investigation, except within a narrow region located at the edge of the contact, which corresponds to the observed crack initiation sites. Such a contact mechanics analysis therefore supports the hypothesis of a potential activation of crack nucleation processes by some localized plastic strain accumulation at the edge of the contact, as a result of the strong strain localization associated with the tangential loading.

A quantitative analysis of contact crack nucleation processes typically involves the following ingredients:

(1) A multiaxial analysis of the cyclic stresses experienced by the material in the vicinity of the crack initiation sites. On the basis of a contact mechanics analysis, the amplitude of the average tensile and shear stresses is calculated along the initial crack direction in order to identify the nature and the level of the local stress field associated to crack initiation.

(2) an estimate of the fatigue properties of the bulk material under loading conditions which are representative of that induced in the initial crack direction. Such a condition is not always easy to fulfill in situations where crack

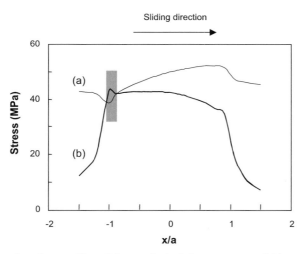

Fig. 11 Calculated surface profiles of the octahedral shear stress at yield assuming a modified Von Mises criterion (*a*), and of the octahedral shear stress for a glass/epoxy contact under gross sliding condition (*b*). The *grey area* delimits the region at the leading edge of the contact where the octahedral shear stress is exceeding the limit octahedral shear stress at yield (*a* is the radius of the contact area) (from [97])

initiation involves a complex multiaxial stress field. However, the analysis detailed below shows that the cracks initiated at the edge of the contacts are induced under predominantly tensile conditions, which facilitates the comparison to the bulk tensile fatigue response.

For elastic materials, the contact problem is usually solved as a unilateral contact problem obeying Coulomb's friction law. The algorithms used here are based on those pioneered by Kalker [66]. The contact area, the stick and slip regions, the pressure and traction distributions are numerically determined first and then the stress and displacement distributions within the elastic bodies can be established at the various stages of the tangential cyclic loading. On the basis of these calculations, the occurrence of crack initiation processes can subsequently be analysed in the meridian plane of the contact, $y = 0$ (Fig. 12), where the cracks first initiate. As a first approach, parameters based on the amplitude of the shear stress, τ_m, along a particular direction and the amplitude of the tensile stress, σ_m, perpendicular to this direction,

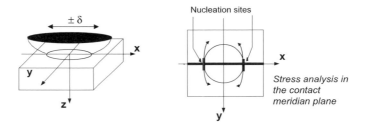

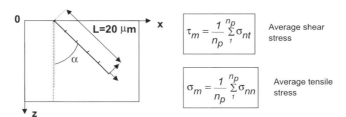

Fig. 12 Theoretical analysis of initial crack growth directions. In a first step, the three-dimensional elastic contact stress field is calculated within the polymer body under small amplitude reciprocating micro-motions. A two-dimensional analysis of crack initiation is subsequently carried out using the calculated stress values in the meridian plane of the contact (Oxz). Average shear (τ_m) and tensile (σ_m) stresses are calculated for different locations in the contact and for different orientations, α, with respect to the normal to the contact plane

can been considered to derive the crack initiation criterion [65]. These mechanical parameters essentially allow discrimination between predominant shear or tensile crack initiation driving forces.

For various discrete steps of the cyclic tangential loading, values of τ_m and σ_m can be calculated for different orientations, α, with respect to the normal to the surface and for different locations within the meridian plane. As an example, for an epoxy specimen under gross slip conditions, the maximum value, $\Delta\sigma_m^*$, of the effective (i.e. $\sigma_m > 0$) amplitude of the average tensile stress on the surface ($z = 0$) has been reported as a function of the orientation in Fig. 13. In the same figure, the orientation, α^*, of the plane corresponding to the maximum amplitude of the effective average tensile stress has also been reported. Owing to the loading symmetry and for the sake of clarity, only the results corresponding to one half of the contact have been represented. The results show that the maximum amplitude of σ_m occurs at the edge of the contact (i.e. $x/a = -1$) and along an orientation, $\alpha^* = 7°$, which is very close to the experimental initial crack propagation direction (11°). The calculations also reveal that, at the edge of the contact, the maximum amplitude of the shear stress, $\Delta\tau_m^*$, is minimized along a direction corresponding to the maximum tensile stress amplitude (Fig. 14). In order to take into account the strong stress gradient close to the contact interface, τ_m and σ_m had to be averaged over a finite length from the polymer surface. The calculated value of α^* was, however, found to be roughly independent of the depth up to 50 µm, while the amplitude of $\Delta\sigma_m^*$ decreased only from 70 MPa to 60 MPa within the same range.

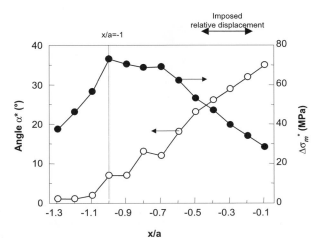

Fig. 13 Calculated amplitude ($\Delta\sigma_m^*$) and orientation (α^*) of the effective average tensile stress in an epoxy polymer as a function of the location within the contact area (gross slip condition, a is the radius of the contact area, and the amplitude of the relative displacement is indicated by the *double arrow*) (from [97])

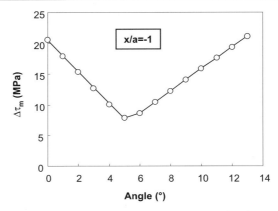

Fig. 14 Calculated values of the maximum shear stress amplitude, $\Delta\tau_m^*$, at the edge of the contact ($x/a = -1$) as a function of the orientation with respect to the normal to the surface (gross slip condition, a is the radius of the contact area) (from [97])

This combined analysis of $\Delta\sigma_m^*$ and $\Delta\tau_m^*$ therefore establishes that the main cracks that nucleate close to the contact edge correspond to predominantly tensile fatigue cracks. This conclusion remains valid whatever the contact condition (partial slip or gross slip). In addition, the distribution of $\Delta\sigma_m^*$ within the contact plane is of interest (Fig. 15). The maximum amplitude

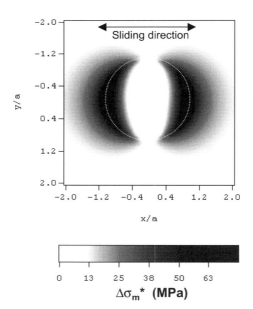

Fig. 15 Calculated distribution of the amplitude of the effective average tensile stress, σ_m^*, within the plane of an epoxy glass contact under gross slip condition

of the tensile stresses is located within two croissant-shaped areas oriented perpendicular to the sliding direction, which correspond closely to the regions where crack initiation was observed experimentally.

The calculated maximum tensile stress amplitude within the contact is of interest in the light of the bulk fatigue data of the DGEBA/IPD network. The latter were obtained under three-point bending conditions and it was verified from observations of the fracture surfaces that cracks nucleated and propagated on the tensile side of the specimen according to a mode I opening mechanism, similar in nature to that involved in contact fatigue. When the calculated value of the tensile stress amplitude at the edge of the contact (i.e. $\Delta\sigma_m^* = 70$ MPa) under gross slip conditions is reported within the corresponding S-N diagram, a fatigue life of about 250 cycles is obtained which corresponds closely to the experimental number of cycles to crack initiation under contact fatigue conditions (Fig. 16). It is worth noting that this good agreement between bulk and contact fatigues behaviours is observed despite the fact that the loading frequencies (10 Hz and 1 Hz for bulk and contact fatigue, respectively) and the strain ratio $R = \varepsilon_{min}/\varepsilon_{max}$ (0.1 and -1.4 for bulk and contact fatigues respectively) are significantly different. Using the same approach, the tensile stress amplitude, $\Delta\sigma_m^*$, can be calculated under partial slip conditions for a displacement amplitude which corresponds to the threshold of crack initiation. The calculated stress values are consistent with the experimental bulk endurance properties of the epoxy material, which fur-

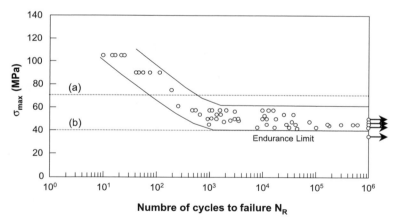

Fig. 16 S-N fatigue diagram of a bulk diglycidyl ether of bisphenol (DGEBA)/isophoron diamine (IPD) epoxy polymer giving the maximum applied stress as a function of the number of cycles to failure (three-point bending, 25 Hz, stress ratio $\sigma_{min}/\sigma_{max} = 0.1$) (from [53]). The two *dotted lines* correspond to theoretical values of the amplitude of the effective tensile stress, $\Delta\sigma_m^*$, calculated for (a) gross slip condition and (b) under partial slip condition for an imposed displacement (± 10 μm) which corresponds to the experimental contact endurance limit at 10^5 cycles

ther supports the relevance of bulk fatigue data within the context of contact fatigue behaviour of brittle epoxy materials.

3.4
Crack Propagation: Ductile-to-Brittle Transitions

The above-described contact mechanics analysis provides a good estimate of the initial crack orientation within the epoxy substrate. The problem of crack orientation is especially relevant in the context of fatigue wear processes because particle detachment is often considered to result from pitting mechanisms associated with the coalescence of adjacent cracks which propagate along different (declined) directions depending on the sliding direction and crack interaction. In such a situation, the wear rate can be assumed to be primarily dependent upon both the crack orientation and crack growth rate.

There is a long history behind the problem of crack propagation within contacts involving brittle materials, starting with the initial studies of Hertz on conical fractures at elastic contacts between curved glass substrates. As detailed in the book by Lawn [67], many of these investigations were in fact motivated by the wide use of indentation fracture as a simple microprobe for determining brittle material fracture parameters such as toughness, crack velocity exponent, etc. Key elements of the related contact problem are stress field inhomogeneity and crack stability, which often result in a breakdown of critical stress concepts to describe the threshold condition for the unstable propagation of cracks. These aspects have been addressed in particular within the context of indentation cracking of brittle material with pre-existing surface flaw distribution. From a combination of an Hertzian contact stress analysis with well known linear elastic fracture mechanics solutions for line cracks, it is possible to derive the values of the stress intensity factors along the trajectories of the Hertzian cones which are often observed under sphere indentation conditions. As an example, the results of such a calculation are reported in Fig. 17, where the stress intensity factors at the crack tip have been normalized with respect to the fracture toughness. Starting from an initial flaw size, c_f, this curve shows that stable (i.e. $dK/dc < 0$, where c is the crack length) and unstable (i.e. $dK/dc > 0$) propagation stages can successively be involved in indentation fracture when the normal load is increased. This approach highlights that the critical load at the threshold for unstable crack propagation is independent on the initial flaw size, but that it is determined by a critical crack depth, c_c, which is achieved after an initial stable propagation stage.

It is worth noting that this calculation of the stress intensity factor does not take into account the modification of the elastic stress field which results from crack propagation. However, the contact stiffness measurements carried out during contact fatigue of epoxy materials (see Fig. 7) show that a substantial change in the magnitude of the tangential loading and in the associated stress

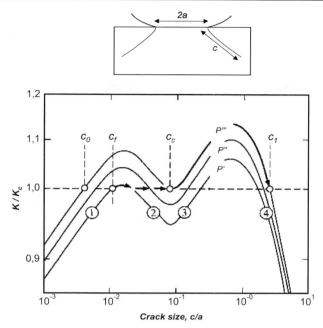

Fig. 17 Contact mechanics analysis of Herztian cracks within brittle materials. **a** Schematic description of a Hertzian cone crack induced under normal indentation by a rigid sphere. **b** Reduced plot of K-field as function of cone crack length and for increasing loads $P' < P'' < P'''$ during sphere-on-flat normal indentation of brittle materials. *Arrowed segments* denote stage of stable ring crack extension from c_f to c_c (initiation), then unstable to c_1 at $P = P'''$ (cone-crack pop-in) (From [67]). *Branches* (1) and (3) correspond to unstable crack propagation ($dK/dc > 0$), *branches* (2) and (4) to stable crack propagation ($dK/dc < 0$)

field can occur when the crack length become of the order of magnitude of the contact radius.

Thus, more recently, refined fracture mechanics approaches have been developed which take into account the modifications of the elastic stress field associated with crack propagation as well as the effects of potential microsliding mechanisms between the crack faces [68–70]. Without going into detail, stress and displacement fields are obtained by superimposing the individual responses of the uncracked solid and of the cracks to the contact loading in a manner that satisfies the boundary conditions along the faces of the cracks The crack response is associated with displacement discontinuities along its faces (opening and slip processes) which generate stresses.

Such approaches have been applied to the contact cracking problems under sliding conditions in order to assess the values of the stress intensity factors K_I and K_{II} under cyclic tangential loading. As an example, Fig. 18 shows the numerical simulations corresponding to a cracked epoxy substrate.

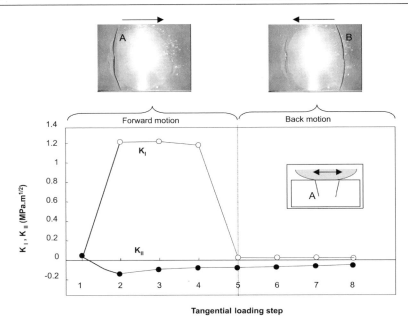

Fig. 18 Calculated values of the mode I (K_I) and mode II (K_{II}) stress intensity factors within a cracked epoxy at various stages of a cyclic contact loading. Two cracks 350 μm in length and oriented at 10° with respect to the normal to the epoxy surface have been considered. For symmetry reasons, only the results corresponding to one crack (denoted A) have been represented. The tangential cyclic loading has been divided into eight successive steps

The crack behaviour and the associated stress intensity factors during the various discrete steps of the cyclic tangential loading are represented. During the alternate tangential loading, opening (steps 1 to 4) and closing (steps 5 to 8) mechanisms of the crack were simulated in accordance with the *in situ* observation of the crack dynamics. This calculation of K_I and K_{II} also clearly demonstrates the complex non-proportional nature of the loading at the crack tip. It is usually found, however, that for such cracks the magnitude of K_{II} is much less than that of K_I, which means that the tip loading is essentially of a mode I nature.

When considering the use of such a fracture mechanics analysis to analyse the propagation of relatively short cracks within epoxy, some complications could arise from the potential formation and breakdown of crazes. Although the existence of crazes in epoxies has been the matter of some debate, there appears to be overwhelming evidence against the occurrence of such failure mechanisms in bulk specimens of these polymers [71]. Investigation of mode I failure mechanisms in epoxy resins also invariably indicates that, depending on the temperature, the loading rate and the yield properties, two distinct type of propagation mode can be encountered, namely a stable (continuous) propagation or an unstable "stick-slip" mode [71–73]. The unstable crack-growth behaviour is generally attributed to localized crack tip yielding [73]. Accordingly, the fracture energies for crack initiation increase with increasing temperatures (or decreasing loading rates) due to the temperature and strain rate dependence of the yield stress. On the other hand, fracture energies for crack arrest were nearly independent of temperature and were similar to the fracture energies for stable propagation. The yield stress therefore appears to be the controlling parameter regarding the nature of crack propagation (Fig. 19). During the course of contact fatigue experiments, in situ observations indicated that the brittle propagation stages observed within the mixed and gross slip regimes proceeded in a stable manner. These observations are consistent with the above analysis and the fact that relatively high yield stresses are expected for the investigated epoxy systems far below T_g and at the high loading rates associated with the cyclic contact loading.

In Fig. 20, the maximum calculated values of K_I are reported as a function of the crack length under a constant amplitude tangential loading. As the crack length increases, the maximum value of K_I is progressively increased up to the experimental value of the bulk fracture toughness ($K_{IC} = 1.2$ MPa m$^{1/2}$ according to [74]). Accordingly, the simulation indicates that a transition from stable to unstable crack propagation should occur during the course of a contact fatigue experiment, which is consistent with the above-reported experimental observations. Taking into account the experimental value of

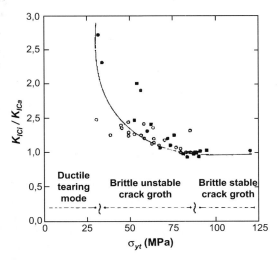

Fig. 19 Correlation between the ratio of the stress intensity factor to crack initatiation (K_{ci}) to stress intensity factor at crack arrest (K_{ca}) and the true yield stress (σ_{yt}) of epoxy materials (from [73])

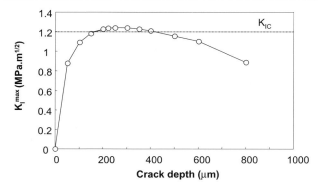

Fig. 20 Calculated maxima in the mode I stress intensity factor as a function of crack depth for a DGEBA/IPD epoxy material under cyclic loading. The *dotted line* corresponds to the experimental value of the mode I fracture toughness

the toughness, the calculation indicates that this transition should occur for a critical crack depth of about 200 µm, which closely matches experimental values.

4
The Role of Molecular Parameters in Contact Fracture Processes of Glassy Polymers

4.1
Toughened Epoxy Networks

The results described above clearly show that the intrinsic low fracture toughness of highly cross-linked epoxy networks can be a severe limitation in tribological applications involving contact fatigue conditions. Some attempts have been made, however, to assess the potential of toughened epoxy systems in order to improve the wear resistance of these materials. The addition of rubbers or thermoplastics which are initially miscible in the epoxy system and display a phase separation during curing (RIPS) has been recognized as an efficient route for toughening epoxies [75, 76]. Carboxyl and amine terminated butadiene acrylonitrile (CTBN and ATBN, respectively) rubbery particles in particular proved to be successful in improving fracture behaviour but at the price of a substantial decrease in the thermo-mechanical properties due to the incorporation of the rubber phase into the epoxy matrix [77, 78]. Moreover, these elastomers have relatively high glass transition temperatures which limit their low temperature flexibility. To overcome this limitation, siloxane elastomers emerged as an attractive alternative because they have glass transition temperatures well below those of CTBN and ATBN. When incorporated into highly cross-linked epoxy networks, poly(dimethylsiloxane) and poly(dimethyl-co-diphenyl) particles were found to induce a substantial increase in the wear resistance [79–81]. The formation of contact cracks within the siloxane-modified epoxies appears, however, to be strongly influenced by the size and the distribution of the micrometre-scale rubber domains which segregate during curing. When the domains are large and closely packed, there is some evidence that particle detachment processes result from the localization of the crack paths at the particle/matrix interface [80]. For the classical pin-on-disk tribological test configurations considered in these investigations, the wear rates roughly correlated with the inverse of the fracture toughness. The precise identification of the contribution of fracture toughness to the observed improvement in wear properties is, however, complicated by the lubricating action of polysiloxane-containing wear debris, which accumulates at the sliding interface during the course of the wear process. As a consequence, the magnitude of the cyclic stresses induced within the substrate is reduced and it is not clear whether the enhancement of the wear properties results from the improvement of the intrinsic fracture properties rather than from the modification of the friction-induced surface tractions.

Some more detailed answers to this question were recently provided within the context of the analysis of the contact fatigue behaviour of anti-plasticized epoxy networks [82]. Anti-plasticizing agents usually consist of small molecule additives that can be incorporated into epoxy networks in order to increase the room temperature modulus. This latter effect has been shown to result from a decrease in the mechanical losses associated with the sub-T_g β relaxation, by virtue of the partial hindrance of the associated molecular motions by the additives molecules [83, 84]. In addition, the selection of specific anti-plasticizing additives with a reduced miscibility in the epoxy resin can also yield an improvement in fracture properties. Bearing in mind that epoxy network toughening systematically results from phase separation, an appropriate tuning of the additive polarity allows selection of molecules that are initially miscible in the monomer mixture, but give rise to nanoscale phase separation during curing. Such a goal was achieved by Sauvant et al. [85, 86] by incorporating acetamide derivates in epoxy formulations based on diglycidylether of bisphenol A (DGEBA) fully cured in the presence of a stoichiometric amount of aromatic diamine (Table 1). In addition to an increased room temperature modulus, these formulations also exhibited a twofold increase in the room temperature fracture toughness (Table 2).

The contact fatigue behaviour of these modified epoxy networks was investigated under small amplitude micromotions. Under similar contact loading conditions, no cracks were detected within an anti-plasticized DGEBA/DDM network up to 5×10^3 cycles, whereas early crack propagation was detected in the unmodified network. Although some slight wear degradation was ob-

Table 1 Formulae of the modified stoichiometric epoxy networks

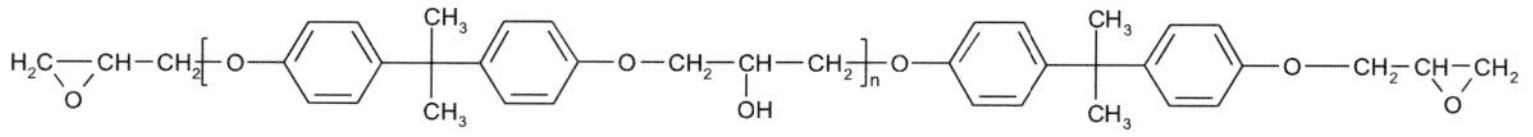

DGEBA

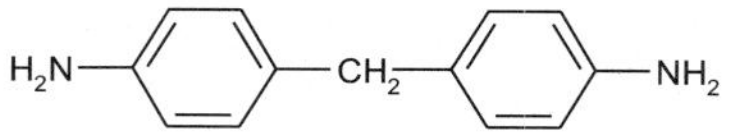

DDM

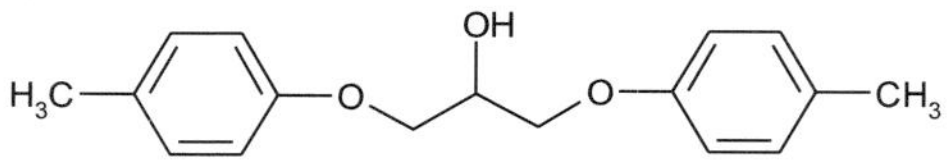

Acetamide derivative (AM)

Table 2 Mechanical and physical properties of the unmodified and modified DGEBA/DDM epoxy polymers. Data taken from [85, 86]. *AM* Acetamide derivative (see Table 1)

	T_g (°C)[a]	E' (MPa)[b]	σ_y (MPa)[c]	K_{IC} (MPa m$^{1/2}$)[d]
DGEBA/DDM	190	2800	100	0.8
DGEBA/DDM-AM	111	3400	110	2.1

[a] Measured from the maximum of the loss modulus E'' at 1 Hz using dynamic mechanical analysis
[b] Conservation modulus measured at 25 °C and 1 Hz
[c] Compressive yield stress measured at a strain rate of 2×10^{-3} s^{-1}
[d] Mode I Fracture toughness from notched three point bending specimens

served within the contact area (Fig. 21), the epoxy network with the included additive can be considered as essentially undamaged at the end of the fretting tests.

A relevant analysis of these differences in terms of toughness requires, however, that the local stress conditions at the crack nucleation sites are properly evaluated. Owing to changes in the modulus and/or frictional behaviour of the modified epoxy material, different stress conditions can be achieved locally in the contact and the improved crack resistance of modified epoxy does not necessarily reflect a change in its intrinsic crack propagation resistance. Regarding the magnitude of the tensile stresses generated at the edge of the contact, two opposite effects came into play: (1) an increase in the contact pressure due to the "fortified" modulus of the anti-plasticized material (from an Hertzian contact analysis, the quasi-static Young's moduli of the

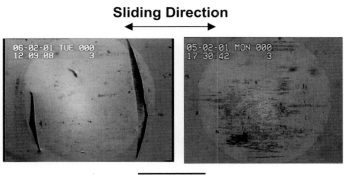

Fig. 21 In situ visualization pictures showing damage **a** in a fretting contact between an unmodified DGEBA/DDM epoxy network and a glass sphere, and **b** in a similar contact with a DGEBA/DDM network modified with the acetamide derivative shown in Table 1. The magnitude of the maximum tensile stress at the edge of the contact was found to be similar in both cases. Contact fatigue cracks were only observed in the neat DGEBA/DDM system

unmodified and modified epoxy networks were found to be 2600 MPa and 3300 MPa respectively), and (2) a slight decrease in the coefficient of friction in the case of the anti-plasticized network (from 1.2 to about 1.1). It turned out that these two effects compensated to give roughly the same value of the maximum tensile stress amplitude at the edge of the contact when crack nucleation was detected. Accordingly, the improved contact fatigue resistance of the anti-plasticized networks can be unambiguously attributed to their intrinsic toughening. Tentatively, a realistic mechanism might involve stress concentration localized in the nanophase domains which is able to favour plastic flow and dissipation at the crack tip, in agreement with the accepted toughening mechanism involved in rubber-toughened epoxy systems.

4.2
Copolymers of Methylmethacrylate

Random copolymers of methylmethacrylate (MMA) also provide a convenient way to investigate the contribution of molecular parameters to the contact fatigue behaviour of glassy amorphous polymers. Extensive mechanical characterization by Halary and co-workers [87–92] showed that the incorporation of various amounts of glutarimide or N-substituted maleimide units as a co-monomer of MMA (Table 3) can result in strong changes in the plastic behaviour and deformation modes. At the molecular level, these modifications were mostly attributed to changes in the cooperative motions involved in the temperature range situated immediately above the sub-T_g β relaxation, which are considered precursors of the α-relaxation [90]. Addition of increas-

Table 3 Formulae of the random copolymers of methylmethacrylate

Methacrylate-glutarimide copolymers (GIM)	[MMA unit and GIM unit structures shown]
Methacrylate-maleimide copolymers (MIM)	[MMA unit and MIM unit structures shown]

ing amounts of maleimide was shown to induce a decoupling of the α and β relaxation, which was associated with a marked embrittlement of the materials. Conversely, increased contents in glutarimide units tended to enhance the cooperative motions involved in yielding with some consequences on the competition between scission crazing, shear deformation and disentanglement crazing in these materials [92]. Homologous series of MMA-glutarimide (GIM) or MMA-maleimide (MIM) copolymers therefore offer the possibility of tuning the fracture properties from a ductile to a brittle behaviour, while keeping the elastic and plastic properties almost unchanged (Table 4).

A systematic investigation of the contact fatigue behaviour of the GIM and MIM copolymers has been carried out at various imposed normal loads within the elastic range of the polymers [93]. Whatever the contact load, a strong difference was observed in the development of the crack networks for the two materials (Fig. 22). For the MIM system, the nucleation of cracks at the edge of the contact area was immediately followed by a brittle propagation stage. In contrast, crack propagation within the GIM copolymer occurred in a much more progressive manner and multiple crack initiation was often observed.

Crack propagation within MIM was also associated with a slight but perceptible decrease in the contact tangential stiffness, which was not the case for the GIM copolymer. As reported above, the contact stiffness is strongly sensitive to the extent of crack propagation. The fact that a drop in contact stiffness is not perceptible for the GIM polymer could therefore be interpreted as evidence of a limited propagation of the surface cracks through the thickness of the specimens.

Table 4 Mechanical and physical properties of the methylmethacrylate-based random copolymers. *GIM 76* is a copolymer containing 76% of glutarimide units, *MIM 25* a copolymer containing 25% maleimide units. Data taken from [89, 90, 98]

	T_g (°C)[a]	E' (MPa)[b]	σ_y (MPa)[c]	K_{IC} (MPa m$^{1/2}$)	H (MPa)
GIM 76	158	3400	120	2.4[d]	260
MIM 25	161	3000	125	0.85[f]	250

[a, b, c, d] As for Table 2
[d] As for Table 2. Measured using nano-indentation at a constant average strain rate of 0.05 s^{-1}. [f] Estimated from the fretting tests (see text)

Interestingly, the ductile–brittle transition observed for the MIM system provided an opportunity to assess the material fracture toughness, which was not possible using classical fracture mechanics tests due to the intrinsic brittleness of the MIM system. The measurement of the critical crack length, L_c, in the contact plane at the onset of brittle propagation allows estimation of a fracture toughness $K_{IC} = \sigma_{xx}^{max}\sqrt{\pi L_c}$ in the order of 0.85 MPa m$^{1/2}$, i.e. much less than that of a poly(methylmethacrylate) homopolymer (1.20 MPa m$^{1/2}$).

As for the epoxy polymers, a quantitative comparison of the contact fatigue behaviour was attempted on the basis of an estimate of the maximum tensile stress at the edge of the contact. The coefficient of friction of the copolymers increased as the tests proceeded, with a variation which was dependent upon the level of the normal loading. As a first approach, the value of μ at crack initiation was taken into account in the calculation of σ_{xx}^{max}. The results are reported in a "S-N" fatigue diagram giving the maximum applied tensile stress as a function of the number of cycles to crack initiation (Fig. 23). These data show a marked increase in the contact fatigue resistance of the GIM copolymers compared with the MIM material.

The observed contact cracking mechanisms are interesting to compare to previously reported investigations of the microdeformation mechanisms of thin films of the same copolymers strained in tension and observed by transmission electron microscopy [90, 92]. Depending upon the temperature and the composition of the methylmethacrylate copolymers, competitive deformation processes in the form of crazes or homogeneous shear deformation zones were identified. For the MIM copolymer under investigation, extensive craze formation was observed at room temperature as a result of the embrittlement of the copolymers by the maleimide co-monomers. In contrast, straining of the GIM copolymer thin films at 20 °C resulted in the extensive formation of shear bands in addition to crazes (as a comparison poly(methylmethacrylate) shows predominantly crazes under comparable conditions). This behaviour reflects the increased plastic yielding ability, or ductility, of the GIM copolymer and has a strong beneficial ef-

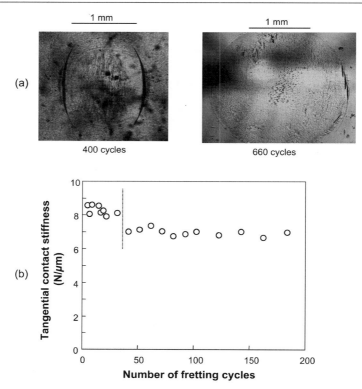

Fig. 22 Contact cracking processes within random copolymers of methylmethacrylate under fatigue conditions (from [93]). **a** In situ observation of the contact areas. **b** Changes in the contact stiffness as a function of the number of cycles for the cyclohexyl maleimide system (see Table 2). Crack propagation is associated with a drop in contact stiffness as indicated by the *dotted line*. In the case of the glutarimide system, no significant change in contact stiffness was detected after contact cracking

fect on the macroscopic fracture toughness, which is about twice that of poly(methylmethacrylate).

These conclusions are intersting in the light of the investigations by Yang et al. on the role of craze breakdown during the fracture processes involved in abrasive wear in glassy polystyrene and blends of polystyrene and poly(phenylene oxide) [39, 40]. Using series of monodisperse polystyrenes with increasing molecular weight, these authors showed a correlation between the craze breakdown strain measured using thin films and the abrasive wear resistance of the polymer. Similar trends were also observed when the craze breakdown strain was modified by incorporating various amounts of low molecular weight polystyrene into the high molecular weight polystyrene material. The use of fully miscible blend systems of craze forming polystyrene and ductile poly(phenylene oxide) also provided the possibility of investigat-

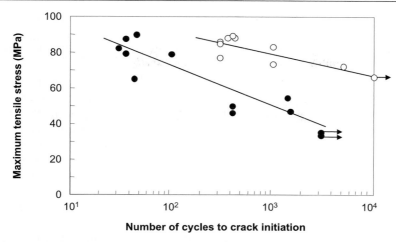

Fig. 23 Contact fatigue behaviour of methylmethacrylate random copolymers under fretting conditions. *White circle* Glutarimide co-monomer (76%), *black circle* cyclohexyl malemimide co-monomer (25%). The maximum value of the tensile stress has been calculated at the edge of the contact, where cracks initiate, and for values of the coefficient of friction measured at crack initiation

ing the effects on wear behaviour of brittle–ductile transitions from crazing to shear yielding. An examination of the wear-composition dependence of these systems showed the existence of two wear regimes delimited by the known composition for crazing-shear yielding transition. Accordingly, Yang et al. concluded that the wear transition resulted from a change from craze-dominated abrasive wear processes at low poly(phenylene oxide) contents to a behaviour where shear zones became the prevalent microdeformation mechanism. These experimental observations tend to support the idea that deformation and fracture modes during asperity scratching of polymeric glasses is strongly controlled by a competition between crazing processes and shear yielding. It also suggests that craze criteria could be adequately be incorporated in a contact mechanics analysis of crack nucleation in PS and acrylate polymers. Interestingly, it can be noted that usual craze nucleation criteria incorporate the effect of hydrostatic pressure [94, 95] which can be shown to be maximum at the leading edge of the contact where cracks nucleate.

5
Conclusion

Extensive in situ observations of failure processes within model single asperity contacts demonstrate the relevance of fracture mechanics approaches to the analysis of cracking processes within sliding contacts involving brittle amorphous polymers. Under predominantly elastic conditions, it turns out that key parameters such as crack location, orientation and propagation depth can adequately be predicted from a knowledge of the bulk material fracture properties (toughness, fatigue) and a detailed analysis of the contact stress field and stress intensity factors. This conclusion tends to support the overall validity of a fracture mechanics approach to the quantitative analysis of initial particles detachment processes involved in the wear of brittle polymers. The incorporation of fracture mechanics ingredients into realistic wear models would naturally require that interactions between multiple cracks are taken into account. On the basis of the available description of single-crack propagation mechanisms, there is, however, no obvious objection to the extension of the current fracture mechanics approach to more realistic situations involving the development of a complete crack network within a sliding contact.

The combination of experimental and theoretical approaches to contact cracking processes was also applied to homologous series of polymers differing in their microstructural characteristics or by the presence of low molecular weight additives. It allowed a clear distinction between the separate contributions of intrinsic toughness properties and frictional properties to the observed changes in the contact cracking behaviour. Such an approach could provide some guidelines for the development of wear-resistant polymer formulations in situations where brittle cracking is the dominant tribological damage mechanism.

One of the outstanding problems related to contact cracking in brittle polymers remains the complex interactions between crack nucleation processes and surface modifications due to frictional sliding. In the above detailed fretting analysis, cracks were indeed induced by the contact stress field, but in a region (at the edge of the contact zone) where the polymer surface was not subjected to extensive frictional processes. The situation would be different under large amplitude sliding in the sense that cracks will be initiated in a surface layer extensively sheared during the frictional processes. Within this context the question of determination of the extent to which surface microstructural modifications due, for example, to cyclic plastic deformation, can affect crack nucleation processes arises. In contrast to metallic materials where the interactions between particle detachment processes and tribologically transformed surface layers has long been recognized, this field of investigation remains largely unexplored in polymer tribology.

References

1. Lancaster JK (1972) In: Jenkins AD (ed) Polymer science, a material science handbook, vol 2. North-Holland, London, pp 959–1042
2. Lancaster JK (1990) Wear 141:159

3. Briscoe BJ, Sinha SK (2002) J Eng Tribol 216:401
4. Briscoe BJ (1990) Scripta Metall Mater 24:839
5. Briscoe BJ (1981) Tribol Int 14:231
6. Briscoe BJ, Chateauminois A (1998) In: Vincent L (ed) Matériaux et contacts. Presses Polytechniques et Universitaires Romandes, Lausanne, pp 197–208
7. Bowden FP, D Tabor (1958) The friction and lubrication of solids. Clarendon, Oxford
8. Bahadur S (2000) Wear 245:92
9. Proceedings of the Royal Society of London - Series A, 1972, 329 (1578), p 251
10. Tanaka K, Miyata T (1977) Wear 41:383
11. Hironaka S (1995) Sekiyu Gakkaishi 38:137
12. Plummer CJG, Kausch H-H (1996) Polym Bull 37:393
13. Plummer CJG, Kausch HH (1997) Sen'i-Gakkaishi 53:555
14. Li TQ, Zhang MQ, Song L, Zeng HM (1999) Polymer 40:4451
15. Ratner SN, Farberoua II, Radyukeuich OV, Lure EG (1964) Soviet Plastics 7:37
16. Lancaster JK (1969) Tribology Convention. Institute of Mechanical Engineers, London p 100
17. Atkins AG, Omar MK, Lancaster JK (1984) J Mater Sci Lett 3:779
18. Jain VK, Bahadur S (1980) Wear 60:237
19. Jain VK, Bahadur S (1982) Wear 79:241
20. Berthier Y, Kapsa P, Vincent L (1998) In: Vincent L (ed) Matériaux et contacts. Presses Polytechniques et Universitaires Romandes, Lausanne
21. Godet M (1984) Wear 100:437
22. Briscoe BJ, Chateauminois A, Lindley TC, Parsonage D (1998) Tribol Int 31:701
23. Briscoe BJ, Chateauminois A, Parsonage D, Lindley TC (2000) Wear 240:27
24. Chateauminois A, Briscoe BJ (2003) Surf & Coatings Technol 163:435
25. Briscoe BJ (1998) Tribol Int 31:121
26. Friedrich K (1998) In: Friedrich K (ed) Advances in composite tribology. (Composite materials series), vol 8. Elsevier, Amsterdam p 209
27. Tewari US, Bijwe J (1993) In: Friedrich K (ed) Advances in composite tribology. (Composite materials series), vol 8. Elsevier, Amsterdam p 159
28. Schottner G, Rose K, Posset U (2003) J Sol-Gel Sci Technol 27:71
29. Stojanovic S, Bauer F, Glasel HJ, Mehnert R (2004) Prog Adv Mater Processes (Mater Sci Forum) 453–454:473
30. Mehnert R, Hartmann E, Glasel HJ, Rummel S, Bauer F, Sobottka A, Elsner C (2001) Materialwissenschaft Werkstofftechnik 32:774
31. Schallamach A (1971) Wear 17:301
32. Tabor D (1951) The Hardness of Metals. Oxford University Press, New York
33. Briscoe BJ, Fiori L, Pelillo E (1998) Journal of Physics D: Applied Physics 31:2395
34. Briscoe BJ, Pelillo E, Sinha SK (1996) Polymer Engineering and Sci 36:2996
35. Briscoe BJ, Evans PD, EP, Sinha SK (1996) Wear 200:137
36. Briscoe BJ, Pelillo E, Sinha SK (1997) Polymer International 43:359
37. Briscoe BJ, Sinha SK (2003) Materialwiss Werkstofftech 34:989
38. Xiang C, Sue HJ, Chu J, Coleman B (2000) J Polym Sci: Part B: Polymer Physics 39:47
39. Yang ACM, Wu TW (1993) J Mater Sci 28:955
40. Yang ACM, Wu TW (1997) J Polym Sci B: Polym Phys 35:1295
41. Jardret V, Morel P (2003) Prog Org Coatings 48:322
42. Hamilton GM (1983) Proc Inst Mech Eng 197C:53
43. Bucaille JL, Felder E, Hochstetter G (2001) Wear 249:422
44. Bucaille JL, Felder E (2002) Philos Mag A-Phys Condensed Matter Struct Defects Mech Properties 82:2003

45. Bertrand-Lambotte P, Loubet JL, Verpy C, Pavan S (2001) Thin Solid Films 398–399:306
46. Bertrand-Lambotte P, Loubet JL, Verpy C, Pavan S (2002) Thin Solid Films 420–421:281
47. Puttick KE (1980) J Phys D: Appl Phys 13:2249
48. Puttick KE, Yousif RH (1983) J Phys D: Appl Phys 16:621
49. Gauthier C, Lafaye S, Schirrer R (2001) Tribol Int 34:469
50. Gauthier C, Schirrer R (2000) J Mater Sci 35:2121
51. Zhang HQ, Sadeghipour K, Baran G (1999) Wear 224:141
52. Sadeghipour K, Baran G, Zhang HQ, Wu W (2003) J Eng Mater Technol Trans Asme 125:97
53. Chateauminois A, Kharrat M, Krichen A (2000) In: Chandrasekaran V, Elliott CB (eds) Fretting fatigue: current technology and practices, ASTM STP 1367. American Society for Testing and Materials, West Conshohocken, p 325
54. Mindlin RD (1953) ASME Trans J Appl Mech, Ser E 16:327
55. Johnson KL (1985) Contact mechanics. Cambridge University Press, Cambridge, UK
56. Kharrat M, Krichen K, Chateauminois A (1999) Tribol Trans 42:377
57. Vincent L (1994) Fretting fatigue ESIS 18. Mechanical Engineering Publications, London, p 323
58. Fouvry S, Kapsa P, Vincent L (1996) Wear 200:186
59. Cattaneo C (1938) Rend Accad Lincei 6:343
60. Nagasawa M, Kinuhata H, Koizuka H, Miyamoto K, Tanaka T, Kishimoto H, Koike T (1995) J Mater Sci 30:1266
61. Lorenzo L, Hahn HT (1986) Polym Eng Sci 26:274
62. Rabinowitz S, Beardmore P (1974) J Mater Sci 9:81
63. Lesser AJ, Kody RS (1997) J Polym Sci B: Polym Phys 35:16
64. Kody RS, Lesser AJ (1997) J Mater Sci 32:5637
65. Dubourg MC, Chateauminois A, Villechaise B (2003) Tribol Int 36:109
66. Kalker JJ (1990) Three dimensional elastic bodies in rolling contact. Kluwer, Dordrecht
67. Lawn B (1993) Fracture of brittle solids, 2nd edn. Cambridge University Press, Cambridge, UK
68. Dubourg MC, Villechaise B (1992) ASME J Tribol 114:462
69. Dubourg MC, Villechaise B (1989) Eur J Mech A 8:309
70. Dubourg MC, Villechaise B (1992) ASME J Tribol 114:455
71. Yamini S, Young RJ (1980) J Mater Sci 15:1823
72. Phillips DC, Scott JM, Jones M (1978) J Mater Sci 13:311
73. Vakil UM, Martin GC (1993) J Mater Sci 28:4442
74. Urbaczewski-Espuche E, Galy J, Gerard JF, Pascault JP, Sautereau H (1991) Polym Eng Sci 31:1572
75. Williams RJJ, Rozenberg B, Pascault JP (1997) Adv Polym Sci 128:1
76. Sultan JN, McGarry FJ (1973) Polym Eng Sci 13:29
77. Becu L, Sautereau H, Maazouz A, Gerard JF, Pabon M, Pichot C (1994) Polym Adv Technol 6:316
78. Verchere D, Pascault JP, Sautereau H, Moschiar SM, Riccardi CC, Williams J (1991) J Appl Polym Sci 42:701
79. Rey L, Poisson N, Maazouz A, Sautereau H (1999) J Mater Sci 34:1775
80. Chitsaz-Zadeh MR, Eiss NS (1990) Tribol Trans 33:499
81. Eiss NS, Czichos H (1986) Wear 111:347
82. Chateauminois A, Sauvant V, Halary JL (2003) Polym Int 52:507

83. Heux L, Lauprêtre F, Halary JL, Monnerie L (1998) Polym 39:1269
84. Merritt ME, Goetz JM, Whitney D, Chang CPP, Heux L, Halary JL, Schaefer J (1998) Macromolecules 31:1214
85. Sauvant V, Halary JL (2001) J Appl Polym Sci 29:759
86. Sauvant V, Halary JL (2002) Composites Sci Technol 62:481
87. Halary JL, Monnerie L, Kausch HH (2005) In: Kausch HH (ed) Intrinsic molecular mobility and toughness of polymers. (Advances in polymer science). Springer, Berlin Heidelberg New York 187:215
88. Laupretre F, Monnerie L, Halary JL (2005) In: Kausch HH (ed) Intrinsic molecular mobility and toughness of polymers. (Advances in polymer science). Springer, Berlin Heidelberg New York 187:35
89. Tordjeman P, Halary JL, Monnerie L, Donald AM (1995) Polymer 36:1627
90. Tordjeman P, Teze L, Halary JL, Monnerie L (1997) Polym Eng Sci 37:1621
91. Teze L, Halary JL, Monnerie L, Canova L (1999) Polymer 40:971
92. Plummer CJG, Kausch H-H, Teze L, Halary JL, Monnerie L (1996) Polymer 37:4299
93. Lamethe JF, Sergot P, Briscoe BJ, Chateauminois A (2003) Wear 255:758
94. Argon AS, Hannoosh JG (1977) Philos Magazine 36:1195
95. Estevez R, Van der Giessen E (2005) In: Kausch HH (ed) Intrinsic mobility and toughness of polymers. (Advances in Polymer Science). Springer, Berlin Heidelberg New York 188:195
96. Briscoe BJ (1998) In: Friedrich K (ed) Advances in composite tribology (composite materials series), vol 8. Elsevier, Amsterdam, p 3
97. Dubourg MC, Chateauminois A (2002) In: Williams JG (ed) Fracture of polymers, composites and adhesives (ESIS TC4) (ESIS publication 32). Elsevier, Les Diablerets, Switzerland, p 51
98. Teze L, Halary JL, Monnerie L, Canova L (1998) Polymer 40:971

Modeling and Computational Analysis of Fracture of Glassy Polymers

R. Estevez[1] (✉) · E. Van der Giessen[2]

[1] GEMPPM-CNRS, INSA Lyon, 20 av Albert Einstein, 69621 Villeurbanne cedex, France
Rafael.Estevez@insa-lyon.fr

[2] Department of Applied Physics, Micromechanics of Materials Group,
University of Groningen, Nyenborgh 4, 9747 AG Groningen, The Netherlands
E.Van.der.Giessen@rug.nl

1	Introduction	197
2	Viscoplastic Deformation of Amorphous Polymers	198
3	Crazing	203
3.1	Craze Initiation	204
3.2	Craze Thickening	205
3.3	Craze Breakdown	207
4	Cohesive Surface Model for Crazing	212
4.1	Craze Initiation	213
4.2	Craze Thickening	214
4.3	Craze Breakdown	214
4.4	Traction versus Thickening Law for the Cohesive Surfaces	215
4.5	Alternative Descriptions	218
5	Computational Analysis of Glassy Polymer Fracture	218
5.1	Problem Formulation	219
5.2	Isothermal Analysis	221
5.3	Coupled Thermomechanical Analysis	227
6	Conclusion and Future Trends	232
	References	233

Abstract Although it is recognized that failure of glassy polymers involves crazing and shear yielding, most of the studies of their fracture account for one *or* the other mechanism. We present a finite element analysis in which crazing *and* shear yielding are incorporated. Shear yielding is accounted for through the description of a three-dimensional constitutive law of the bulk material, while crazing is modeled by a cohesive surface which comprises the three stages of initiation, thickening, and craze fibril breakdown and related crack formation. The description is able to capture the main features of glassy polymer fracture such as the ductile-to-brittle transition at low rates and the evolution of the toughness with loading rate. In particular, it is demonstrated that the competition between shear yielding and crazing governs the material's toughness. Even if the description of crazing presented here is essentially phenomenological, a cohesive zone formulation is shown to provide a consistent formulation to bridge descriptions of failure at the molecular length scale with analyses performed at the continuum scale.

Keywords Crack tip plasticity · Elastic–viscoplastic material · Crazing · Cohesive surface · Fracture

Abbreviations

$\otimes$	Tensor product
$\cdot$	Scalar product
$(\ )'$	Deviatoric part of a second-order tensor
tr	Trace of a tensor
e_i	Basis unit vector
I	Second-order unit tensor
D	Deformation rate tensor
D^e	Elastic part of the deformation rate tensor
D^p	Plastic part of the deformation rate tensor
σ	Cauchy stress tensor
b	Back stress tensor
$\bar{\sigma}$	Effective stress tensor
π, T	Second Piola–Kirchhoff stress tensor and related traction vector
η, u	Lagrangian strain tensor and displacement vector
$\overset{\triangledown}{\sigma}$	Jaumann rate of the Cauchy stress tensor
C_e	Fourth-order isotropic elastic modulus tensor
T_g	Glass transition temperature
E	Young's modulus
K	Bulk modulus
ν	Poisson's coefficient
α_c	Coefficient of cubic thermal expansion
$\dot{\gamma}^p$	Equivalent shear strain rate
τ	Equivalent shear stress
p	Hydrostatic pressure
α	Pressure sensitivity coefficient
s_0	Initial shear strength
s_{ss}	Steady-state shear strength
h	Parameter controlling the rate of softening
$\dot{\gamma}_0$	Pre-exponential shear strain rate factor
A	Parameter for the temperature dependence of the shear strain rate
$\dot{\mathcal{D}}$	Energy dissipation rate per unit volume
λ_α	Principal plastic stretches ($\alpha = 1, ..., 3$)
$\bar{\lambda}$	Largest plastic stretch
e_α^p	Principal directions of the plastic stretch
$b_\alpha^{3\text{-ch}}, b_\alpha^{8\text{-ch}}$	Three chains and eight chains description for the estimate of the back stress tensor b
λ_N	Maximum stretch of the polymer coil
N	Number of segments between entanglements
n	Volume density of entanglements
k_B	Boltzmann constant
C^R	Shear modulus of the plastically activated entangled network
R	Gas constant
E_a	Entanglement dissociation energy
η	Viscosity in the non-Newtonian response above T_g
m	Exponent of the non-Newtonian response above T_g

k	Thermal conductivity
ρ	Mass density
c_v	Isochoric specific heat
σ_y	Tensile yield stress
σ_m	Mean stress
I_1	First stress invariant
$\sigma_1, \sigma_2, \sigma_3$	Principal stresses
$\sigma_{max}, \sigma_{min}$	Maximum and minimum principal stresses
σ_b	Stress bias
A^0, B^0	Parameters in the craze initiation criterion of Sternstein et al. [24]
X', Y'	Parameters in the craze initiation criterion of Oxborough and Bowden [26]
Δ^{cr}	Critical craze thickness
Λ_c	Craze length
D_0, D	Primitive and mature fibril diameter
V_e	Craze thickening rate
τ_0	Fibril lifetime
σ_c	Craze stress in the Dugdale description (see Figs. 3 and 4)
f_b	Force for chain breakage
v_s	Surface density of entangled chains
E_1, E_2	Young's moduli to represent the anisotropic craze fibrils structure
σ_n	Traction vector normal to the cohesive surface
σ_n^{cr}	Critical traction for craze initiation
σ_t	Traction vector tangential to the cohesive surface
$\dot{\Delta}_n^c$	Craze thickening rate along the direction normal to the craze plane
$\dot{\Delta}_0, A^c, \sigma^c$	Material parameters for the normal craze thickening rate: preexponential thickening rate, temperature dependence, athermal normal stress
$\dot{\Delta}_t^c$	Tangential craze thickening rate
$\dot{\Gamma}_0, \tau^c$	Material parameters for the tangential craze thickening rate: pre-exponential thickening rate, athermal shear stress
k_n	Stiffness of the cohesive surface
h_0	Thickness of the craze at the onset of craze initiation
q	Heat flux through the cohesive surfaces
G_c	Energy release rate
K_I	Mode I stress intensity factor
$\dot{K}_I$	Rate of the mode I stress intensity factor
M_w	Weight-average molecular weight
PMMA	Poly(methyl methacrylate)
PC	Polycarbonate
SAN	Styrene–acrylonitrile polymer

1
Introduction

Amorphous polymers exhibit two mechanisms of localized plasticity: crazing and shear yielding. These are generally thought of separately, with crazing corresponding to a brittle response while shear yielding is associated with ductile behavior and the development of noticeable plastic deformation prior to fracture. Shear yielding is plastic flow localized in a shear band caused by the inherent strain softening after yield followed by rehardening at continued deformation. Crazing is also a mechanism of localized plasticity but at a different scale, and has a distinctly different appearance. Crazes are planar cracklike defects but unlike cracks, the craze surfaces are bridged with polymer fibrils resulting in some load-bearing ability. Following the presently available description of crazing, we present how it can be modeled by cohesive surfaces. The related constitutive response is written in terms of a traction-opening law which incorporates the stress-state dependence for initiation, the rate-dependent thickening during fibrillation, and the fibril breakdown for a critical craze thickness.

Most of the analyses of crazing found in the literature so far assume that the bulk surrounding the craze remains elastic. This assumption is certainly a limitation to the analysis of glassy polymer fracture, since crazing and shear yielding can appear simultaneously [1]. As both are rate-dependent processes, the competition between their kinetics for a given loading (defined by its level and rate) is expected to control which mechanism is developing first and dominates, thus defining a ductile or brittle response.

We present recent results on the analysis of the interaction between plasticity and crazing at the tip of a preexisting crack under mode I loading conditions. Illustrations of the competition between these mechanisms are obtained from a finite element model in which a cohesive surface is laid out in front of the crack.

As the loading rate increases, thermal effects need to be accounted for and the analysis is extended to a coupled thermomechanical framework. Evidence of a temperature effect in glassy polymer fracture is found (e.g., in [2, 3]) with a temperature increase beyond the glass transition temperature T_g. The influence of thermal effects on the fracture process is also reported.

Tensors are denoted by bold-face symbols, $\otimes$ is the tensor product, and $\cdot$ the scalar product. For example, with respect to a Cartesian basis e_i, $\boldsymbol{AB} = A_{ik}B_{kj}\boldsymbol{e}_i \otimes \boldsymbol{e}_j$, $\boldsymbol{A} \cdot \boldsymbol{B} = A_{ij}B_{ij}$, and $\boldsymbol{C}^\mathrm{e}\boldsymbol{B} = C^\mathrm{e}_{ijkl}B_{kl}\boldsymbol{e}_i \otimes \boldsymbol{e}_j$, with summation implied over repeated Latin indices. The summation convention is not used for repeated Greek indices.

2
Viscoplastic Deformation of Amorphous Polymers

We present a constitutive model for amorphous polymers in their glassy state ($T < T_\mathrm{g}$) when no crazing takes place (like in shear or in compression). The formulation is supplemented with a simple description of the material response when the temperature gets higher than T_g, as found experimentally to occur at sufficiently high loading rates [2, 3]. Therefore, two descriptions of the viscoplastic response of amorphous polymers are used, depending on the temperature. The two viscoplastic processes are described within the same framework.

Following the original ideas of Haward and Thackray [4], the three-dimensional basis for modeling the deformation of glassy polymers is due to Boyce et al. [5]. The constitutive model is based on the formulation of Boyce et al. [5], but we use a modified version introduced by Wu and Van der Giessen [6]. Details of the governing equations and the computational aspects were presented by Wu and Van der Giessen in [7]. The reader is also referred to the review by Van der Giessen [8], together with a presentation of the thermomechanical framework in [9].

The constitutive model makes use of the decomposition of the rate of deformation D into an elastic, D^e, and a plastic part, D^p, as $D = D^e + D^p$. Prior to yielding, no plasticity takes place and $D^p = 0$. In this regime, most amorphous polymers exhibit viscoelastic effects, but these are neglected here since we are primarily interested in those of the bulk plasticity. Assuming the elastic strains and the temperature differences (relative to a reference temperature T_0) remain small, the thermoelastic part of the response is expressed by the hypoelastic law

$$\overset{\triangledown}{\sigma} = C_e D^e - K\alpha_c \dot{T} I, \tag{1}$$

where $\overset{\triangledown}{\sigma}$ is the Jaumann rate of the Cauchy stress, and C_e the usual fourth-order isotropic elastic modulus tensor. The coefficients K and α_c are the bulk modulus and the coefficient of cubic thermal expansion, respectively. Assuming that the yield response is isotropic, the isochoric plastic strain rate D^p is given by the flow rule

$$D^p = \frac{\dot{\gamma}^p}{\sqrt{2}\tau} \bar{\sigma}', \tag{2}$$

which is specified in terms of the equivalent shear strain rate $\dot{\gamma}^p = \sqrt{D^p \cdot D^p}$, the driving stress $\bar{\sigma} = \sigma - b$, and the related equivalent shear stress $\tau = \sqrt{1/2\, \bar{\sigma}' \cdot \bar{\sigma}'}$. The back stress tensor b describes the progressive hardening of the material as the strain increases and will be defined later on.

The equivalent shear strain rate $\dot{\gamma}^p$ is taken from Argon's expression [10]

$$\dot{\gamma}^p = \dot{\gamma}_0 \exp\left[-\frac{As_0}{T}\left\{1 - \left(\frac{\tau}{s_0}\right)^{5/6}\right\}\right] \quad \text{for } T < T_g, \tag{3}$$

where $\dot{\gamma}_0$ and A are material parameters and T the absolute temperature (note that plastic flow is inherently temperature dependent through Eq. 3). In Eq. 3 the shear strength s_0 is related to elastic molecular properties in Argon's original formulation but is considered here as a separate material parameter. In order to account for the effect of strain softening and for the pressure dependence of the plastic strain rate, s_0 in Eq. 3 is replaced by $s + \alpha p$, where α is a pressure sensitivity coefficient and $-p = 1/3 \mathrm{tr}(\sigma)$. Boyce et al. [5] have suggested a modification of Eq. 3 to account for intrinsic softening by substituting s_0 with s, which evolves from the initial value s_0 to a steady-state value s_{ss} according to $\dot{s} = h(1 - s/s_{ss})\dot{\gamma}^p$, with h controlling the rate of softening. The energy dissipation rate per unit volume is given by

$$\dot{\mathcal{D}} = \bar{\sigma}' \cdot D^p = \sqrt{2}\tau\dot{\gamma}^p. \qquad (4)$$

The resulting temperature rise will be accounted for in the coupled analysis to be presented in Sect. 5.

The constitutive model is completed by the description of the progressive hardening of amorphous polymers upon yielding due to the deformation-induced stretch of the molecular chains. This effect is incorporated through the back stress b in the driving shear stress τ in Eq. 2. Its description is based on the analogy with the stretching of the cross-linked network in rubber elasticity, but with the cross-links in rubber being replaced with the physical entanglements in a flowing amorphous glassy polymer [5]. The deformation of the resulting network is assumed to be affine with the accumulated plastic stretch [6], so that the principal back stress components b_α are functions of the principal plastic stretches λ_β as

$$b = \sum_\alpha b_\alpha (e_\alpha^p \otimes e_\alpha^p), \quad b_\alpha = b_\alpha(\lambda_\beta), \qquad (5)$$

in which e_α^p are the principal directions of the plastic stretch. In a description of the fully three-dimensional orientation distribution of non-Gaussian molecular chains, Wu and Van der Giessen [6] showed that b can be estimated accurately with the following combination of the classical three-chain model and the eight-chain description of Arruda and Boyce [11]:

$$b_\alpha = (1 - \xi) b_\alpha^{3\text{-ch}} + \xi b_\alpha^{8\text{-ch}}, \qquad (6)$$

where the fraction $\xi = 0.85\bar{\lambda}/\sqrt{N}$ is based on the maximum plastic stretch $\bar{\lambda} = \max(\lambda_1, \lambda_2, \lambda_3)$ and on N, the number of segments between entanglements. The use of Langevin statistics for calculating b_α implies a limit stretch of $\sqrt{N}$. The expressions for the principal components of $b_\alpha^{3\text{-ch}}$ and $b_\alpha^{8\text{-ch}}$ contain a second material parameter, the initial shear modulus $C^R = nk_BT$, in which n is the volume density of entanglements (k_B is the Boltzmann constant).

Based on a study of the temperature dependence of strain-induced birefringence in amorphous polymers, Raha and Bowden [12] suggested that the thermal dissociation of entanglements can be described by

$$n(T) = B - D \exp(-E_a/RT) \qquad (7)$$

In the above expression, E_a is the dissociation energy, R is the gas constant, and B and D are material parameters. Such evolution of the entanglement density is used to model a reduction of the hardening with temperature. As pointed out by Arruda et al. [13], this evolution law is subject to the side condition $nN = $ constant in order to keep the number of molecular links constant. Therefore, the back stress according to Eq. 5 is also temperature dependent through $N(T)$ and $C^R = n(T)k_B T$. The material parameters B and D are estimated here from the assumption that the back stress vanishes as the temperature approaches T_g, resulting in $n(T_g) = 0$ so that $B/D = \exp(-E_a/RT_g)$.

The formulation above is assumed to hold for temperatures up to the glass transition T_g. For $T > T_g$, most studies found in the literature focus on the description of the molten state [14] due to its practical importance, while little attention is paid to the response of glassy polymers in the rubbery state, near T_g. For strain rates larger than 1 s^{-1}, the mechanical response of the molten material is non-Newtonian for most polymers and described by $\tau = \eta \dot{\gamma}^m$, where η and m are material parameters. We assume that this non-Newtonian response prevails as soon as T_g is exceeded. Hence, within the same framework as used below T_g, the equivalent plastic strain rate (Eq. 3) is replaced by

$$\dot{\gamma}^P = \dot{\gamma}_0 \aleph \left(\frac{\tau}{s_0}\right)^{1/m} \quad \text{for } T \geq T_g. \tag{8}$$

In this expression, η as been substituted for convenience by $\eta = s_0/(\aleph \dot{\gamma}_0)^m$ with s_0 and $\dot{\gamma}_0$ being below-T_g parameters in Eq. 3, and $\aleph$ a nondimensional constant. The deformation in the molten state is generally believed to involve chain slippage and temporary entanglements between the moving chains resulting in the non-Newtonian viscosity [14]. The details of the deformation process are lumped into the parameters η (or $\aleph$) and m so that no back stress contribution is considered above T_g: $\bar{\sigma} \equiv \sigma$ and $\tau = \sqrt{1/2\,\sigma\cdot\sigma}$.

The exponent m is observed to vary between 0.3 and 1 for molten polymers [14] but for those exhibiting a marked non-Newtonian response like most glassy polymers, m ranges from 0.3 to 0.5; the value $m = 0.4$ is adopted here. For a given temperature, the evolution of the viscosity η with increasing strain rate is observed to decrease from a Newtonian value η_0 at low strain rates to a level five or six decades smaller [14, 15]. For a temperature around T_g, the value of η_0 can be estimated from [15] to be of the order of some megapascals for materials like PMMA or PC. A smaller value is expected in the non-Newtonian regime so that for temperatures above T_g, a constant value of $\eta = 0.35$ MPa ($\aleph = 0.02$) is used to describe the material response in the molten state. With this simple description, we only aim at being able to continue our calculations if the temperature exceeds T_g locally, which may happen during crack propagation at high loading rates. We need to keep in mind that we are primarily concerned with temperatures below T_g, and the incorporation of more sophisticated models as found in [14, 15] is out of the scope of the present investigation.

Figure 1 shows the response to simple shear that is obtained with the constitutive model described above under isothermal and adiabatic conditions,

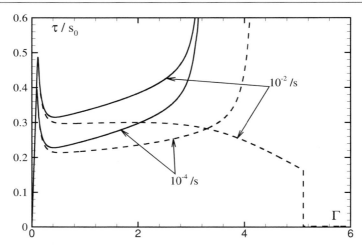

Fig. 1 Mechanical response of SAN to simple shear at an applied strain rate of 10^{-2}/s and 10^{-4}/s for isothermal or adiabatic conditions (*solid* and *dashed lines*, respectively). The temperature increase delays the hardening for a strain rate of 10^{-4}/s which vanishes for 10^{-2}/s. In this case, the material reaches the glass transition temperature and enters the rubbery state resulting in a small load-bearing capacity

i.e., when the viscoplastic energy dissipation converted into heat as

$$\rho c_v \dot{T} = \dot{\mathcal{D}}, \tag{9}$$

in which $\dot{\mathcal{D}}$ is defined in Eq. 4, ρ is the mass density, and c_v the specific heat. The temperature increase reduces the rehardening due to the back stress tensor b since the entanglement density n decreases with increasing temperature (see Eq. 7), thus resulting in a higher maximum stretch $\sqrt{N}$ of the molecular strands.

The material parameters used are given in Table 1 and are representative of SAN. The parameters involved in adiabatic analyses (Eq. 9) are reported in Table 2 and are expected to be representative for glassy polymers. For the lowest shear strain rate of 10^{-4}/s, adiabatic conditions moderately affect the orientational hardening which appears for a slightly larger deformation. As the loading rate is about 10^{-2}/s or higher, the final hardening is suppressed

Table 1 The set of parameters used for the description of the bulk response representative of SAN at room temperature supplemented with $\aleph$ and m involved in the description for $T > T_g$

	E/s_0	ν	s_{ss}/s_0	As_0/T	h/s_0	α	N	C^R/s_0	$\aleph$	m
SAN	12.6	0.38	0.79	52.2	12.6	0.25	12.0	0.033	0.02	0.4

Table 2 The set of parameters used in the thermal part of the analysis from [9] and standard for glassy polymers

E_a/R	k	α_c	ρ	c_v
2.8×10^3 K	0.35 W/mK	2×10^{-4} K^{-1}	1.08×10^{-3} kg/m^3	1.38×10^3 J/kg K

and the glass transition temperature is reached locally, corresponding to the loss of any load-bearing capacity of the material. It is unlikely that adiabatic conditions are met for a strain rate of 10^{-2}/s but as long as temperature effects need to be considered, this figure shows that the material may not harden, even for large deformations.

3
Crazing

Since early investigations of glassy polymer fracture, as reviewed in [16], crazing has been recognized as the mechanism preceding the nucleation of a crack. The major advances in the description of the crazing process are thoroughly examined in two volumes of the present series [18, 19]. The mechanism of crazing involves three stages (see [16–19]): (1) initiation, (2) thickening of the craze surfaces, and (3) breakdown of the craze fibrils and creation of a crack. The description presented here is phenomenological since the model is motivated by the mechanical considerations, and the molecular aspects (e.g., the molecular weight or the entanglement density) are generally not incorporated.

Because crazing is a precursor to failure, pioneering studies on crazing have focused on the conditions for craze initiation. Later on, estimation of the toughness motivated examination of the craze thickening and the conditions for craze breakdown.

We present the major results established in the description of crazing and the recent developments in this field. Crazing has been investigated within continuum or discrete approaches (e.g., spring networks or molecular dynamics calculations to model the craze fibrils), which have provided phenomenological or physically based descriptions. Both are included in the presentation of the crazing process, since they will provide the basis for the recent cohesive surface model used to represent crazing in a finite element analysis [20–22].

3.1
Craze Initiation

The physical mechanism for craze initiation is not yet clearly identified, and various criteria have been proposed depending on the assumed mechanism and length scale for its description. Experimentally, one observes an incubation time for craze formation when a constant stress smaller than approximately half of the yield stress is applied. The total number of craze nuclei reaches a saturation value which increases with applied stress [16, 23]. For these stress levels, the incubation time can be greater than 100 s and this time decreases with increasing stress. Above half of the yield stress, the incubation time becomes negligible [23]. Based on these observations and borrowing some ideas for ductile failure in metals, Argon and Hannoosh [23] developed a sophisticated criterion for time-dependent craze initiation, which includes a negligible influence of time when the stress level is larger than half the yield stress. In this case the criterion reduces to

$$|\sigma_{max} - \sigma_{min}| = \frac{A}{C + 3\sigma_m/2\sigma_y Q} \tag{10}$$

in which σ_{max} and σ_{min} refer to the maximum and minimum principal stresses, $\sigma_m = 1/3\text{tr}\,\sigma$ is the mean stress, A and C are material parameters, $Q = 0.0133$ is a factor controlling the dependence of the critical shear stress on the mean stress, and σ_y is the tensile yield stress of the material [23]. It is shown in [20] that the criterion (Eq. 10) shows predictions for craze initiation under tension similar to those provided by a formulation based on stress bias conditions as used by Sternstein [24, 25].

The stress bias criterion [24, 25] refers implicitly to two mechanisms of microvoid formation in a dilatational stress field and stabilization of the microvoids through a deviatoric stress component and local plasticity. Its definition is

$$\sigma_b = |\sigma_1 - \sigma_2| \geq A^0 + \frac{B^0}{I_1}, \tag{11}$$

in which σ_b is a stress bias depending on the first stress invariant I_1. Crazing initiates for a positive I_1 and the plane of craze initiation is perpendicular to the direction of maximum principal stress σ_1. Sternstein et al. [24] derive the above expression from experiments in which crazing initiates in the vicinity of a hole drilled in a thin plate of PMMA, for which the principal stresses are $\sigma_1 > \sigma_2 > \sigma_3 = 0$. In a subsequent analysis on thin cylinders under combined tension and torsion loadings, the above formulation is observed to agree with experimental observations of craze initiation when $\sigma_1 > \sigma_3 = 0 > \sigma_2$, with σ_2 being the smaller stress in this case. This is pointed out by Oxborough and Bowden [26], who suggested a definition of a critical strain as $\varepsilon_c \geq X + Y/\sigma_m$, which is also hydrostatic stress dependent. This criterion can be reformulated for an elastic material with ν being the Poisson's coefficient and E the Young's modulus as

$$\sigma_1 - \nu\sigma_2 - \nu\sigma_3 \geq X' + \frac{Y'}{I_1}, \qquad (12)$$

where $X' = EX$ and $Y' = EY$.

The criteria (Eqs. 11 and 12) are similar and are derived from studies on materials that are elastic at initiation of crazing, while more ductile materials like polycarbonate show a more pronounced sensitivity to the hydrostatic tension. This has been found experimentally by Ishikawa and coworkers [1, 27] for notched specimens of polycarbonate. Crazing appears ahead of the notch root, at the intersection of well-developed shear bands. From a slip line field analysis, the tip of the plastic zone corresponds to the location of the maximum hydrostatic stress. This has been confirmed by Lai and Van der Giessen [8] with a more realistic material constitutive law. Therefore, Ishikawa and coworkers [1, 27] suggested the use of a criterion for initiation based on a critical hydrostatic stress. Such a stress state condition can be expressed by Eq. 11 with $\sigma_b = 0$ and $I_1^{cr} = B^0/A^0$. Thus, the criterion (Eq. 11) can be considered general enough to describe craze initiation in many glassy polymers. For the case of polycarbonate, a similar criterion is proposed in [28] as

$$\sigma_1^{cr} = A + \frac{B}{I_1} \qquad (13)$$

in which σ_1^{cr} is the maximum principal stress, and A and B are material parameters. The criterion (Eq. 13) is indeed very similar to Eq. 11 or Eq. 12. As long as a better fundamental understanding of craze initiation is pending, the choice of which criterion to adopt is essentially dependent on the ability of capturing experimental results.

3.2
Craze Thickening

Descriptions of craze thickening are based on the observed crazes at the tip of a stationary crack for creep tests [29, 30] and on observations of crazes in thin films by transmission electron microscopy (TEM) or small-angle X-ray scattering (SAXS) [31, 32].

Döll and coworkers [29, 30] used interferometry to measure the evolution of craze length and craze thickness with time under constant load. By using the Dugdale plastic zone [33] concept for the description of the craze at a stationary crack tip, the craze thickness and the craze length were observed to increase slightly up to 10^5 s and markedly for larger loading times [29, 30]. The craze stress acting on the craze surfaces derived from the Dugdale analysis was observed to decrease with time while thickening continued [29, 30], thus suggesting that crazing is a time-dependent process. Before details of the craze microstructure provided by TEM [34] or SAXS [35] were available, extension of craze fibrils was thought to proceed by creep [16]. If the latter mechanism is operating during craze thickening, it would result in a variation of the craze density along the craze length, since the craze material just after initiation is expected to have a higher density than that in regions with longer craze fibrils. Such a variation of the volume fraction of the craze is not seen in TEM observations of crazes in thin films [31, 32]. Instead, it has been found that the craze microstructure consists approximately of cylinders with a diameter $D \approx 5\text{--}15$ nm for "mature" fibrils, while those at the craze tip are thought to have an initial diameter of $D_0 \approx 20\text{--}30$ nm, depending on the material [31, 32]. Following Kramer [31, 32], the diameter D_0 of the primitive fibrils is assumed to be approximately the fibril spacing in Fig. 2.

These observations appear to be in contradiction with a creep mechanism for craze fibrillation, and the currently accepted description refers to the drawing-in mechanism due to Kramer [31, 32]. Kramer argued that fibrillation takes place within a thin layer (about 50 nm) at the craze/bulk interface, in which the polymer deforms into highly stretched fibrils similar to the mechanism of drawing of polymer fibers, as illustrated in Fig. 2. Craze thick-

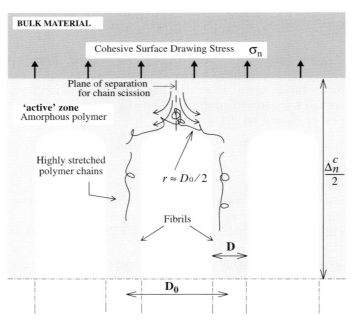

Fig. 2 Description of the craze thickening process according to Kramer [31] as drawing in new polymer chains from the craze/bulk interface into the fibrils. The fibrils have a diameter D and spacing of D_0

ening is then a consequence of large viscoplastic deformations inside this "active" plastic zone at the craze/bulk interface. During fibrillation, stretching of the polymer chains is thought to be combined with chain scission and disentanglement along a plane of separation located at the top of the craze void. Craze fibrils are not only made of parallel cylinders perpendicular to the craze surfaces, but also of lateral cross-tie fibrils which connect the main fibrils. Kramer and Berger [32] suggested that these cross-tie fibrils originate at the plane of separation, when disentanglement is not complete so that fibrillation results in a chain that belongs to two main fibrils.

The deformation of the polymer within a thin active zone was originally represented by a non-Newtonian fluid [31] from which a craze thickening rate is thought to be governed by the pressure gradient between the fibrils and the bulk [31, 32]. A preliminary finite element analysis of the fibrillation process, which uses a more realistic material constitutive law [36], is not fully consistent with this analysis. In particular, chain scission is more likely to occur at the top of the fibrils where the stress concentrates rather than at the top of the craze void as suggested in [32]. A mechanism of local cavitation can also be invoked for cross-tie generation [37].

More work on a detailed description of the fibrillation process is needed to clarify the underlying mechanism and its relationship with molecular aspects, such as the entanglement density or the molecular mobility. Nevertheless, based on the observations reported by Döll [29, 30] of time-dependent craze stress and Kramer's [31, 32] description of fibrillation involving an active plastic zone, one can conclude that craze thickening is a viscoplastic process.

3.3
Craze Breakdown

Following the studies on craze initiation, several efforts have focused on the description of glassy polymer fracture, and especially on the characteristics of a craze developed at a crack tip. Kambour [16] has shown that the length and thickness of a craze developed at the tip of a preexisting crack can be measured by interferometry and quantitative predictions have been reported in [29, 30, 38].

The measure of the craze shape ahead of a propagating crack by Brown and Ward [38] appears consistent with the geometry of the "plastic" zone according to a Dugdale [33] model of a craze. For a precracked specimen under the remote load σ_∞ (Fig. 3), the craze is represented by a plastic zone similar to a strip at the tip of the crack. The profile of the plastic zone varies from zero at the location $(a + \Lambda_c)$ to the value Δ^{cr} at the crack tip.

Typical values for crazes in glassy polymers [29, 30, 38] are a few microns in thickness and tenths of millimeters in length. The measures of Δ^{cr} and Λ_c are used by Brown and Ward [38] to estimate the toughness and the craze

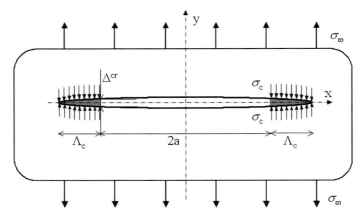

Fig. 3 Schematic of the Dugdale model. The plastic zone is modeled by a strip subjected to a constant normal stress σ_c. The length of the crack is $2a$ and the size of the plastic zone ahead of the crack tip is Λ_c

stress. On the other hand, the observation of a constant critical thickness Δ^{cr} at the tip of a running crack [29, 30, 38] has motivated the definition of a fracture criterion based on this parameter for crack propagation (e.g., in [39, 40]). In PMMA with a high molecular weight ($M_w = 2 \times 10^6$ g/mol), the measured critical craze thickness [41] at the tip of moving cracks is observed to be approximately constant for crack velocities ranging from 10^{-8} mm/s to 20 mm/s, so that the crack velocity moderately affects the critical craze opening.

Depending on the material, a critical molecular weight for the observation of a stable craze has been found, for PMMA [29, 30] and for PC [42]. Below this critical value, crazes are not seen by interferometry and the material is very brittle. The molecular weight has to be sufficiently large (about $M_w = 3 \times 10^5$ g/mol for PMMA and $M_w = 12 \times 10^3$ g/mol for PC) for the development of a stable craze. The critical craze thickness and craze length (Δ^{cr} and Λ_c) are also temperature dependent [29, 30, 43, 44] and this effect is amplified with increasing molecular weight [29, 30].

Based on the description of craze thickening due to Kramer et al. [31, 32], Schirrer [45] proposed a phenomenological viscoplastic formulation for the "fibril drawing velocity" similar to the Eyring model as

$$V_e = V_{e0} \exp\left(\frac{\sigma_c}{\sigma_V}\right), \qquad (14)$$

in which σ_c is the craze stress and σ_V a reference stress (so that $1/\sigma_V$ refers to some activation volume). The maximum craze thickness is obtained by integration of the thickening velocity in Eq. 14 up to a critical time or "lifetime" for the fibrils inside the craze. The lifetime is determined experimentally as

$\tau_0 = \Lambda_c/\dot{a}$, in which Λ_c is the craze length and $\dot{a}$ the crack velocity. This parameter is assumed to be stress dependent as

$$\tau_0 = \tau_{0i} \exp\left(\frac{-\sigma_c}{\sigma_t}\right), \tag{15}$$

similar to that of the fibril drawing velocity in Eq. 14. From the lifetime τ_0 in Eq. 15 and the fibril drawing velocity (Eq. 14), the craze maximum thickness is [45]

$$\Delta^{cr}(\sigma_c) = 2V_{e0}\tau_{0i} \exp(\sigma_c/\sigma_V - \sigma_c/\sigma_t) \tag{16}$$

for a constant craze stress. As the critical craze thickness results from the product of the craze thickening rate (Eq. 14) and the lifetime (Eq. 15), we can notice that if the craze stress σ_c is rate independent, the expression of Δ^{cr} in Eq. 16 is constant, while a rate-dependent craze stress results in a rate-dependent Δ^{cr}, as long as σ_V and σ_t are different. For PMMA, Schirrer [45] indicates that these quantities are very close to each other so that a constant critical craze thickness is derived from Eq. 16. The origin of the craze fibril breakdown is lumped into the definition of the critical craze thickness or in the lifetime τ_0 due to Schirrer [45], the latter formulation including some rate dependence in Δ^{cr} through Eq. 16.

The more recent developments in the description of crazing have addressed the physical origin of the critical craze thickness Δ^{cr} and the features governing its value. Kramer and Berger [32] proposed that craze thickening continues until the entanglement reduction during fibrillation is critically enhanced by the presence of a flaw or a dust particle. Thus, the load-bearing ability locally vanishes and results in a stress concentration. The local increase of the stress triggers the breakdown of the surrounding fibrils and crack propagation within the craze strip. This process involves a statistical analysis presented in [32] but this interpretation neither refers to intrinsic properties nor does it account for the influence of cross-tie fibrils.

Brown [46] demonstrated the importance of the fibrils interconnecting the main fibrils, and a two-scale analysis of the craze strip is considered as presented in Fig. 4. In a Dugdale description (Fig. 4a), the craze zone is subjected to the constant stress σ_c and crack propagation occurs for the critical thickness Δ^{cr}. At the smaller scale represented in Fig. 4b, the craze zone is regarded as a very long strip of thickness δ^f containing a crack. This strip is assumed to be subjected to the remote uniform stress σ_c in the region of the craze–crack transition. The analysis of this local problem aims to estimate the stress distribution within this craze zone and the conditions for crack propagation in terms of strip thickness δ^f, and critical stress for fibril breakdown.

The craze is modeled by an elastic anisotropic medium with Young's moduli E_1 and E_2 corresponding to the stiffness of the cross-tie and the main fibrils ($E_1 \ll E_2$). As the strip thickens along the direction 2 (Fig. 4b), the elastic energy density W is stored so that the energy release rate at failure

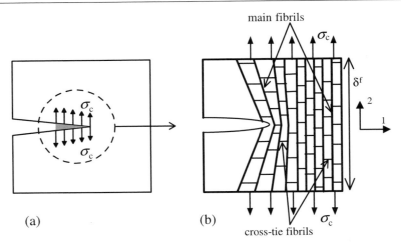

Fig. 4 Description of a craze (**a**) with a Dugdale zone, and the local analysis (**b**) as a long strip representing the anisotropic craze structure made of main fibrils oriented along direction 2, connected with lateral cross-tie fibrils along direction 1

is [46]

$$g = \delta^f W = \delta^f \frac{\sigma_c^2}{2E_2}. \tag{17}$$

Brown suggested the use of Eq. 17 together with the energy release rate estimated for an edge-loaded elastic anisotropic specimen as [46, 47]

$$g \approx \frac{K_I^2}{\sqrt{E_1 E_2}} = \delta^f \frac{\sigma_c^2}{2E_2} \tag{18}$$

to provide an estimate of the stress intensity factor K_I of the local problem. By considering the main fibrils approximately as cylinders of diameter D [31, 32], the estimation of the stress at distance $D/2$ from the crack tip is $\sigma_{22}(D/2) = K_I/\sqrt{\pi D}$ so that the effective stress acting on the closest fibril to the crack tip is

$$\sigma_f = \lambda \sigma_{22}(D/2) = \lambda \sigma_c \left(\delta^f / 2\pi D \right)^{1/2} (E_1/E_2)^{1/4}, \tag{19}$$

in which λ accounts for the isochoric transformation between primordial fibrils to mature fibrils, which results in a reduction of the fibrils' cross-sectional area [31, 32] (the fibril volume fraction is $v_f = 1/\lambda$). The condition for craze fibril breakdown corresponds to $\sigma_f^{cr} = \lambda v_s f_b$ in which v_s is the surface density of entangled chains and f_b the force required for chain breakage. From Eq. 19, the corresponding critical thickness δ^{fcr} for fibril breakdown is estimated from the material parameters (λ, v_s, f_b). The connection between the two analyses comes from the thickness of the fibrillated structure δ^f, the thickness of the initial uncrazed strip δ_0, and the corresponding displacement of the craze surface Δ^c of the Dugdale problem as [31, 34, 46] $\delta^f = \lambda \delta_0 = \Delta^c + \delta_0$. At the onset of craze fibril breakdown, the latter relationship yields $\Delta^{cr} = \delta^{f cr}(1 - v_f)$, in which Δ^{cr} is the critical craze opening of the Dugdale problem and $\delta^{f cr}$ the critical thickness of the craze strip. The energy release rate of the Dugdale problem in Fig. 4a is

$$G_c = \sigma_c \Delta^{cr} = \frac{v_s^2 f_b^2 2\pi D}{\sigma_c} \left(\frac{E_2}{E_1}\right)^{1/2} (1 - v_f), \qquad (20)$$

which is distinct from the approximation used in Eq. 18 for the local problem which only aims to provide an estimate of the thickness δ^f and the stress on the fibril at the crack tip.

The toughness G_c (Eq. 20) is related to material features such as the surface density of entangled chains in the fibrils v_s and the force for chain scission f_b. The value of v_s depends on the amount of disentanglement induced by the fibrillation together with the "initial" entanglement density of the bulk.

Therefore, the parameter v_s provides an interpretation of the molecular dependence of the fracture toughness with molecular weight reported by Döll [29]: fibrillation involves chain scission which can result in a vanishing v_s for an already low molecular weight material to a constant value when stable fibrils are observed. The correlation between the toughness and the entanglement density as predicted by Brown in Eq. 20 has been observed experimentally by Wu [48]. The relationship between the craze stress, the parameters of the craze microstructure (E_1, E_2, and D), and molecular aspects of the polymer chain (flexibility, type of side group) are intensively discussed in the two reviews by Monnerie et al. in this volume, and molecular dynamics calculations devoted to this investigation are emerging [49, 50].

Following Brown's analysis of the influence of the cross-tie fibrils on craze breakdown, several improvements at a length scale between standard continuum mechanics and molecular dynamics have been reported. Hui et al. [51] investigated the strip problem presented in Fig. 4b with different remote conditions: the uniform craze stress σ_c is considered ahead of the crack tip only ($x_1 > 0$ with origin at crack tip) and $\sigma_c = 0$ for $x_1 < 0$; and also a peak stress is used to represent a stress singularity at the crack/craze interface. Predictions of the stress level acting on the fibrils at the crack tip are provided, but the general trends in terms of entanglement surface density and force for chain breakage are similar to those of Brown [46]. Sha et al. [52] use a discrete network of springs to model the fibrils of the craze strip. As two types of springs are used to represent the main or the cross-tie fibrils, estimates of the anisotropic moduli assumed by Brown [46] in terms of the stiffness, the diameter, and the volume fraction of the fibrils are presented.

More recently, Sha et al. [53] pointed out the necessity of using a rate-dependent drawing stress σ_c as reported in [29, 30]. If only this dependence is included in Brown's model, the toughness in Eq. 20 would decrease with increasing craze stress, which is in contradiction with Döll's data [29, 30]. Therefore, Sha et al. [53] incorporated a rate-dependent critical craze thickness in order to capture the evolution of the toughness with crack velocity. However, the evolution of the critical craze thickness predicted in [53] decreases by a factor of five from low to fast crack velocities which is not fully consistent with experimental observations [29, 30], thus indicating that the process of craze fibril breakdown needs to be further clarified.

In conclusion, cross-tie fibrils are important for the interpretation of the mechanism of fibril breakdown, and can explain the influence of the chain entanglement density and chain breakage on the toughness. The analysis of craze breakdown has also pointed out the need for a rate-dependent craze stress, as was already concluded from the craze thickening process. The critical craze thickness is very probably rate dependent as well, but the origin of this still has to be elucidated. Its value is approximately constant for a given temperature and molecular weight, and shows little variation (less than 20%) in PMMA for crack velocities varying over nine decades.

4
Cohesive Surface Model for Crazing

In glassy polymers, crazes have typical dimensions of microns in thickness to tenths of millimeters in length, so that one can generally neglect the craze thickness compared to the other relevant dimensions in the problem under consideration. Following the concept of a cohesive zone due to Needleman [54], one can replace a craze by a cohesive surface, with constitutive properties that are based on the foregoing observations on crazing. Tijssens et al. [20] designed a cohesive surface which mimics the three stages of initiation, thickening, and breakdown. The methodology is illustrated in Fig. 5 with (a) the assumed craze structure, (b) the idealization of the craze process with the transition from initiation to craze thickening and ultimately to craze fibril breakdown and related crack nucleation, and (c) the description in terms of cohesive elements within a finite element framework. When crazing has not yet initiated, the cohesive surfaces are adjacent and there is no discontinuity across the plane under consideration. Once crazing has nucleated, craze thickening develops and the separation between the two cohesive surfaces results in an opening Δ_n. The traction vector σ_n is energetically conjugate to Δ_n and the mechanisms underlying this process are lumped into a traction-opening law to be defined in the sequel. Once craze fibrils break down, a crack nucleates locally and this is accounted for by prescribing a vanishing traction on the corresponding location of the cohesive surface.

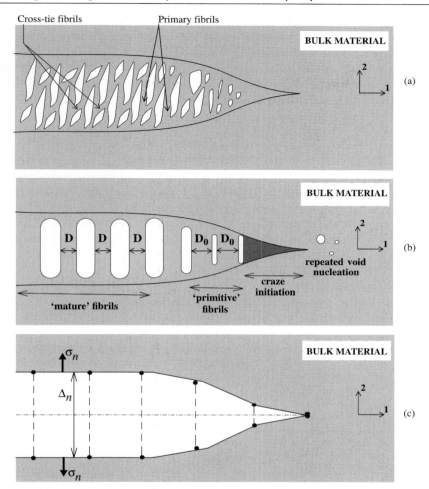

Fig. 5 a Schematic description of the craze structure. **b** Idealization of the craze process according to Kramer and Berger [32] for the craze thickening after initiation. **c** Representation of crazes by discrete cohesive surfaces

4.1
Craze Initiation

Initiation is assumed to occur when a critical stress state is attained, according to one of the criteria reported in the literature as discussed in a preceding section. We choose to use the criterion formulated by Sternstein and Ongchin [24] to illustrate how it can be incorporated into the cohesive surface framework, which is flexible enough to account for another definition. The criterion is presented in Eq. 11 for a plane stress condition, with σ_1 the major principal stress and the first invariant, or $3\sigma_m$. Within the cohesive surface description as in Fig. 5c, we assume that the direction of the major principal stress corresponds to the normal direction so that $\sigma_1 \equiv \sigma_n$ in Eq. 11.

By assuming plane strain conditions as relevant for crack studies and taking the hydrostatic stress as $\sigma_m = 1/3(1 + \nu)(\sigma_1 + \sigma_2) \approx (\sigma_1 + \sigma_2)/2$, the criterion can be reformulated as [22]

$$\sigma_n \geq \sigma_m - \frac{A^0}{2} + \frac{B^0}{6\sigma_m} = \sigma_n^{cr}(\sigma_m). \tag{21}$$

Since σ_1 is the major principal stress, we have $\sigma_n = \sigma_1 \geq \sigma_m = (\sigma_1 + \sigma_2)/2$ and the side condition that the normal stress has to exceed the hydrostatic stress for craze initiation is satisfied. Equation 21 defines a critical normal stress which appears to be hydrostatic stress dependent. As long as $\sigma_n < \sigma_n^{cr}(\sigma_m)$, crazing does not occur and when σ_n reaches $\sigma_n^{cr}(\sigma_m)$ crazing initiates. Once initiated, the craze thickens and the condition (Eq. 21) is no longer relevant.

4.2
Craze Thickening

As discussed in Sect. 3.2, once a mature fibril is created, further thickening occurs by a viscoplastic drawing mechanism which involves intense plastic deformation at the craze/bulk interface [32]. Instead of using a non-Newtonian formulation as in [32] or a formulation based on Eyring's model [45], but on the basis of a preliminary study of the process [36], the craze thickening is described with a similar expression as the viscoplastic strain rate for the bulk in Eq. 3 as [20]

$$\dot{\Delta}_n^c = \dot{\Delta}_0 \exp\left[\frac{-A^c \sigma^c}{T}\left(1 - \frac{\sigma_n}{\sigma^c}\right)\right], \tag{22}$$

in which $\dot{\Delta}_n^c$ is the craze thickening rate and $\dot{\Delta}_0$, A^c, and σ^c are material parameters; $\dot{\Delta}_0$ characterizes the intrinsic mobility during the thickening process, A^c controls the temperature dependence, and σ^c is the athermal stress for craze thickening. The above expression is phenomenological and will be shown to be capable of capturing the main features of glassy polymer fracture. Of course, a physically based formulation would be preferable and could be incorporated when available.

4.3
Craze Breakdown

Craze breakdown is experimentally characterized by a critical craze thickness Δ^{cr} which is primarily dependent (Eq. 20) on the craze stress σ_c, the force for chain scission f_b, and the entangled chain density along the craze surface ν_s. The craze stress σ_c is assumed to be rate and temperature dependent (Sect. 4.2). As the entanglement density before fibrillation is temperature dependent (Eq. 7), the parameter ν_s probably reveals temperature dependence. Experimental observations [29, 30] indicate that the temperature can significantly influence Δ^{cr} but the rate dependence can be neglected as a first approximation.

Therefore, the condition for craze breakdown is incorporated in the cohesive zone description by adopting a critical thickness Δ^{cr} that is just material dependent. We will briefly explore the influence of a temperature-dependent critical thickness in some nonisothermal calculations (Sect. 5.3) by letting the value of Δ^{cr} increase by a factor of two from room temperature to T_g.

Such a phenomenological definition of the critical craze thickness Δ^{cr} hides much of the underlying physics. Further insight is expected to reveal how this parameter changes with loading rate as well as its temperature dependence, which could then be incorporated in the present framework.

4.4
Traction versus Thickening Law for the Cohesive Surfaces

The three stages of the crazing process discussed in the foregoing subsections are combined by the traction-opening law

$$\dot{\sigma}_n = k_n \left(\dot{\Delta}_n - \dot{\Delta}_n^c \right) = k_n \dot{\Delta}_n^e, \tag{23}$$

with $\dot{\Delta}_n$ the normal opening rate of the cohesive surface, $\dot{\Delta}_n^c$ the thickening rate of the craze according to Eq. 22, and k_n the elastic stiffness. The traction-opening law in Eq. 23 is used for the three stages of the crazing process. Prior to craze initiation, $\dot{\Delta}_n^c$ is not relevant and Eq. 23 reduces to $\dot{\sigma}_n = k_n \dot{\Delta}_n$, in which the stiffness k_n has to be "infinitely" large to ensure that the elastic opening remains small and does not significantly affect the continuity of the fields.

When craze widening takes place, k_n represents the elastic stiffness of the fibrillated craze structure. From experimental observations of crazes in thin films, Kramer and Berger [32] suggest that the early stages of fibrillation consist of the transformation of a primitive fibril into a mature fibril at a constant volume. Further fibrillation is due to the drawing process. By considering the primitive fibrils at craze initiation (Fig. 5b) as struts of diameter D_0 and height h_0, their transformation into a mature fibril of diameter D with height h corresponds to $h = \lambda h_0$ and $D = D_0/\lambda^{1/2}$, λ being the extension ratio. By assuming that $\Delta_n = h_0$ at craze initiation, the stiffness of the primitive fibrils is [22]

$$k_n^0 = \frac{\sigma_n^{cr}}{h_0}. \tag{24}$$

During this transformation from primitive to mature fibril, the force distribution acting on the craze/bulk interface remains constant so that $\sigma_n^{f0} = F/S_0$

and $\sigma_n^f = F/S$, in which σ_n^{f0} and σ_n^f represent the stress acting on the primitive and the elongated fibrils, respectively. By assuming that the elastic modulus of the fibrils, E_f, remains constant, we obtain $\sigma_n^{f0} = E_f \Delta_n / h_0$ and $\sigma_n^f = E_f \Delta_n / h$, from which we define the stiffness of the primitive fibrils at craze initiation and that of the mature fibril as

$$k_n^0 = \frac{E_f}{h_0} = \frac{\sigma_n^{cr}}{h_0} \quad \text{and} \quad k_n = \frac{k_n^0}{\lambda}, \tag{25}$$

with $\lambda = h/h_0$. Once the mature fibrils have formed, the craze consists of highly stretched coils. The overall instantaneous elastic stiffness of the craze, k_n, is therefore assumed to arise primarily from the material freshly drawn into the fibrils. The stiffness is thus assumed to remain constant and equal to the limiting value $k_n = k_n^0/\lambda_N$ according to Eq. 25, in which $\lambda_N = \sqrt{N}$ is the maximum stretch of the polymer coil.

Prior to craze initiation, $\sigma_n < \sigma_n^{cr}(\sigma_m)$ in Eq. 21 and the stiffness has to be "infinitely" large to ensure that "no" separation occurs across the cohesive surface. We propose to use

$$k_n^\infty = \frac{\sigma_n^{cr}(\sigma_m)}{h_0} \tag{26}$$

in the traction-opening law (Eq. 23) for numerical convenience, in which σ_m is the instantaneous hydrostatic stress. The stiffness k_n^∞ in Eq. 26 becomes infinite when the mean stress vanishes and a limiting value is used when this happens, which is adjusted to about ten times the stiffness at craze initiation (Eq. 24). When crazing initiates, the stiffness evaluated from Eq. 26 and that from Eq. 24 are identical with k_n^∞ prior to initiation, decreasing gradually to k_n^0.

We illustrate in Fig. 6 the full traction response to a constant widening rate $\dot{\Delta}_n$ derived from Eq. 23. The three regimes of the craze process can be readily distinguished:

1. During the loading, the normal stress σ_n increases but crazing has not yet initiated so that negligible thickening is observed.
2. After a short transition following craze initiation, the craze thickens and results in an opening Δ_n of the cohesive surface at approximately constant normal stress.
3. When the craze thickness attains the critical value Δ_n^{cr}, craze fibrils break down and a microcrack nucleates with a related vanishing normal stress.

During craze thickening, two trajectories are distinguished in Fig. 6, depending on the prescribed widening rate $\dot{\Delta}_n$ of the cohesive surfaces. After initiation, if the cohesive surface widening rate corresponds to $\dot{\Delta}_n > \dot{\Delta}_n^c$, the hardening-like response (2a) is observed while a softening-like response (2b) corresponds to $\dot{\Delta}_n < \dot{\Delta}_n^c$. The plateau is attained when the two opening rates are equal ($\dot{\Delta}_n = \dot{\Delta}_n^c$). The widening rate of the cohesive surfaces $\dot{\Delta}_n$

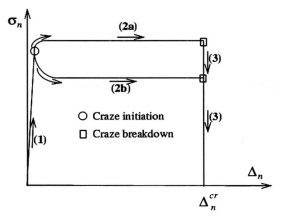

Fig. 6 Schematic representation of the cohesive surface traction-opening law: (1) no crazing, (2) craze widening with (2a) hardening-like response or (2b) softening-like response depending on the prescribed opening rate, and (3) craze breakdown at $\Delta_n = \Delta_n^{cr}$

is governed by the prescribed loading rate, while the craze thickening rate is determined by the craze fibrillation process (Eq. 22). The evolution of the traction-opening law as (2a) or (2b) is controlled by the competition between these two responses. As the area under the traction-opening curve corresponds to the energy dissipated by the crazing process, it also corresponds to the energy release rate for crack propagation. With this formulation, the rate dependence of the energy release rate arises from the rate dependence of the craze thickening process.

The foregoing formulation for crazing focuses on the description along the normal direction. The formulation of Tijssens et al. [20, 21] also accounts for a tangential component in the cohesive properties with a tangential traction σ_t and a tangential displacement Δ_t being related by

$$\dot{\Delta}_t^c = \dot{\Gamma}_0 \left\{ \exp\left[-\frac{A^c \tau^c}{T} \left(1 - \frac{\sigma_t}{\tau^c}\right) \right] - \exp\left[-\frac{A^c \tau^c}{T} \left(1 + \frac{\sigma_t}{\tau^c}\right) \right] \right\}, \quad (27)$$

during drawing ($\dot{\Gamma}_0$ and τ^c are material parameters). From a numerical point of view, an initially large tangential stiffness is necessary to prevent tangential sliding along the cohesive surfaces. The aim in [20, 21] is to use this phenomenological description of the tangential deformation of the cohesive surface to investigate how a variation in the tangential load-carrying ability of the craze fibrils could affect crack propagation and the related resistance curve. This is shown to be of minor importance in [21], but the framework is flexible enough to incorporate such an aspect when an appropriate description is available.

4.5
Alternative Descriptions

Besides the cohesive surface framework presented here, other approaches to model crazing within a finite element formulation have been reported recently. Boyce et al. [55] define a thin craze element which accounts for craze initiation and craze thickening. Craze breakdown was not incorporated, since it was not of concern in the problem under investigation. In many aspects, this formulation resembles a cohesive surface development, but a standard continuum description is adopted which results in a stress–strain relation instead of a traction-opening law. Before crazing, the craze initiates in the craze (continuum) element, and the related thin layer has the properties of the bulk. When the condition for craze initiation is fulfilled within the layer, the thin bulk transforms into a primordial craze. The description of craze thickening uses also a rate-dependent formulation for the normal and the tangential components of the craze thickening, but the analysis is restricted to stable crazes and craze breakdown is not considered.

Gearing and Anand [28] have also used a classical continuum approach to describe crazing. A viscoplastic stress–strain law is derived from an analysis of the material response in the presence of diffuse crazing. This can be justified as long as multiple crazing is observed, but becomes questionable when the loading conditions result in the development of a single craze as in mode I crack growth.

The cohesive surface description presented here has some similarities to the thermal decohesion model of Leevers [56], which is based on a modified strip model to account for thermal effects, but a constant craze stress is assumed. Leevers focuses on dynamic fracture. The thermal decohesion model assumes that heat generated during the widening of the strip diffuses into the surrounding bulk and that decohesion happens when the melt temperature is reached over a critical length. This critical length is identified as the molecular chain contour.

5
Computational Analysis of Glassy Polymer Fracture

We present a finite element study which includes both shear yielding and crazing within a finite strain description. This provides a way of putting together all aspects of glassy polymer fracture: crazing and shear yielding but also thermal effects.

5.1
Problem Formulation

Lai and Van der Giessen [8] performed a finite element analysis of the crack tip plasticity of a mode I crack in glassy polymers, without accounting for crazing. They showed that due to the particular softening–rehardening of glassy polymers, plasticity develops in the form of shear bands. Plastic incompatibility at the shear bands' intersection results in an enhanced hydrostatic stress, the maximum of which is located along the crack symmetry plane. Since hydrostatic stress promotes craze initiation, as indicated in Eq. 11 or Eq. 12, it is expected that crazing appears preferentially along the crack symmetry plane.

Subsequently, Estevez et al. [22] incorporated a cohesive surface along the crack symmetry plane to permit the development of crazing together with bulk plasticity. They assumed bulk plasticity to remain confined near the crack tip so that the small-scale yielding framework is allowed. A single cohesive surface is used, but the framework allows cohesive surfaces to be embedded throughout the volume as performed in [20, 21].

The boundary layer approach is used to investigate the mode I plane strain fields near the crack. The symmetry of the problem allows consideration of only half the geometry (see Fig. 7), which consists of an initial blunted crack of radius r_t with traction-free surfaces along the crack. Along the boundary of the remote region at a distance R with $R \approx 200 r_t$, the mode I elastic field at stress intensity factor K_I is prescribed [8, 22].

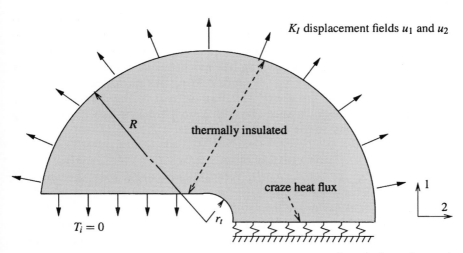

Fig. 7 The small-scale yielding problem in which crazing is allowed along the crack symmetry plane with a plastic bulk. The boundary conditions of the coupled thermomechanical problem are reported from [22, 57]

A quasistatic finite strain analysis is based on the rate form of the principle of virtual work,

$$\Delta t \int_V \left(\pi^{ij} \delta\eta_{ij} + \pi^{ik} u^j_{,k} \delta u_{j,i} \right) dV + \Delta t \int_{S_c} \dot{\sigma}_n \delta \Delta_n dS \quad (28)$$

$$= \Delta t \int_{\partial V} \dot{T}^i \delta u_i dS - \left[\int_V \pi^{ij} \delta\eta_{ij} dV + \int_{S_c} \sigma_n \delta \Delta_n dS - \int_{\partial V} T^i \delta u_i dS \right]$$

in which V and ∂V denote the volume of the region in the initial configuration and its boundary, and where S_c is the cohesive surface in the current state. In Eq. 28, π is the second Piola–Kirchhoff stress tensor, T the corresponding traction vector, and η and u are the conjugate Lagrangian strain and displacement vectors, respectively. The governing equations are solved for each increment, with the term within the square brackets being an equilibrium correction [7, 20].

Basu and Van der Giessen [9] extended the above isothermal formulation to account for thermal effects associated with the heat dissipated by plastic dissipation of the bulk and from the craze process. The plastic energy dissipation rate per unit volume is specified in Eq. 4, so that the energy balance inside the material can be written as

$$\rho c_v \dot{T} = k \nabla^2 T + \bar{\sigma} \cdot D^p, \quad (29)$$

with k the isotropic heat conductivity in accordance with Fourier's law, c_v the specific heat, and ρ the mass density.

There is a second source of energy dissipation, namely the fibrillation process during craze thickening [57]. Per unit of area, the dissipation amounts to $\sigma_n \dot{\Delta}_n^c$ and represents a heat flux $q = -k\nabla T$ into the system through the surface of the cohesive zone. If crazing has not initiated, there is no heat flux across the symmetry plane $x_2 = 0$. Once crazing has nucleated, craze thickening takes place and the above heat flux normal to the craze surfaces is considered. The energy balance (Eq. 29) is then subjected to the following boundary conditions on the cohesive surfaces:

$$\frac{\partial T}{\partial x_2} = \begin{cases} 0 & \text{on } x_2 = 0, \text{without a craze} \\ \frac{1}{2}\sigma_n \dot{\Delta}_n^c / k & \text{on } x_2 = \pm \frac{1}{2}\Delta_n(x_1), \text{along the craze surfaces,} \end{cases} \quad (30)$$

and all others boundaries are insulated.

The heat equation (Eq. 29) is coupled to the equations governing the mechanical response through the temperature dependence of the bulk viscoplastic strain rate (Eq. 3), the craze thickening rate (Eq. 22), and the thermal expansion in Eq. 1. The system of differential equations resulting from the finite element discretization of the energy balance in [9] is modified [57] to include the heat flux vector $\mathcal{D}_c$ from the crazing process as follows:

$$E\dot{\Theta} + F\Theta = \mathcal{D}_b + \mathcal{D}_c \qquad (31)$$

in which Θ is the vector of nodal temperatures and $\mathcal{D}_b$ is the heat source vector due to plastic dissipation in the bulk. The matrices E and F depend on the properties (ρ, c_v), and k, respectively [9]. Equation 31 is integrated in time by an unconditionally stable central difference scheme and the same finite element mesh is used as for the mechanical part (Eq. 28). The coupled problem is handled in a staggered manner as in [9].

5.2
Isothermal Analysis

To illustrate the influence of the craze thickening kinetics on fracture, two sets of craze parameters are used and listed in Table 3. The two sets are borrowed from [22] (cases 8 and 1) and named here A and B. In Fig. 8, we report the plastic strain rate distribution observed near the crack tip. This variable is suitable to track the development of plasticity and is normalized with $\dot{\Gamma}_0 = \dot{K}_I/s_0\sqrt{r_t}$ as a reference strain rate at the tip of the notch (the radius is $r_t = 0.1$ mm and $T = 293$ K). We compare the cases for which no crazing is considered (Fig. 8a) to those where crazing is accounted for (Fig. 8b), with the set A of craze parameters in Table 3. When crazing is not present and at the particular loading rate considered, plasticity develops in the form of shear bands which originate from the tip of the notch, where the stress concentrates.

Figure 8b shows the plastic strain rate distribution at the same load level but with crazing accounted for and developing along the crack symmetry plane. For this case, the craze initiates at the notch root and propagates forward. The comparison between the two distributions shows that crazing reduces the amount of plastic deformation necessary to accommodate the same loading. The distribution shown in Fig. 8b corresponds to the instant prior to the first fibril breakdown. Once this starts, a crack is quickly formed inside the craze and rapid crack propagation takes place, so that the load at this stage can be considered as the critical stress intensity factor. This observation of the craze profile indicates that craze breakdown initiates at the

Table 3 The sets of craze parameters used in this study; the only difference originates from A^c and hence the sets exhibit different craze thickening kinetics (from [22])

Set	A^0/s_0	$B^0/(s_0)^2$	Δ_n^{cr}/r_t	σ^c/s_0	$A^c\sigma^c/T$	$\dot{\Delta}_0/r_t(s^{-1})$
A	0.68	1.4	0.1	0.83	136.5	100
B	0.68	1.4	0.1	0.83	44.6	100

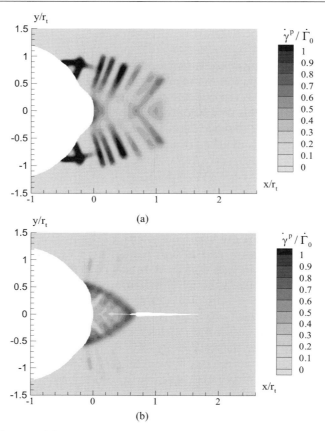

Fig. 8 Distribution of the instantaneous plastic strain rate distribution $\dot{\gamma}^p$ under mode I loading at $\dot{K}_I \approx 3 \times 10^{-2}$ MPa$\sqrt{m}$/s when $K_I/\left(s_0\sqrt{r_t}\right) \approx 1.71$; **a** without crazing, **b** with crazing

point where currently most active shear bands intersect, thus providing an example of how the competition or interaction between shear yielding and crazing can take place. Since both shear yielding and crazing are viscoplastic processes, these mechanisms are also interacting or competing with the loading rate. This is illustrated in Fig. 9 in which the same problem is examined but at a loading rate 120 times higher. In the absence of crazing (Fig. 9a), we observe that shear yielding is more localized than in Fig. 8a. Since the loading rate has increased, the viscoplastic process in the bulk requires a higher loading level for accommodation to be possible; hence, for a comparable loading level, the plasticity appears to be reduced with increasing loading rate. This effect is still observed in the presence of crazing, but the rate dependence of craze thickening also needs to be considered. As a consequence of both processes, the plasticity developed prior to the onset of fibril breakdown

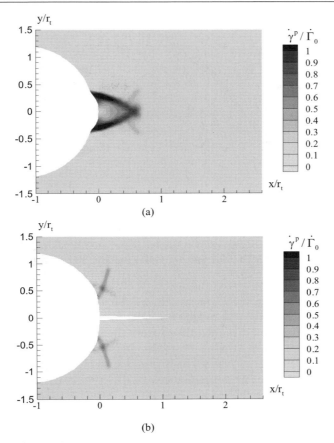

Fig. 9 Distribution of the instantaneous plastic strain rate distribution $\dot{\gamma}^p$ under mode I loading at $\dot{K}_I \approx 3.8$ MPa$\sqrt{m}$/s when $K_I/\left(s_0\sqrt{r_t}\right) \approx 1.56$; **a** without crazing, **b** with crazing

in Fig. 9b is significantly reduced. This also affects the craze profile which is thicker at the tip of the notch in this case, and leads to the crack nucleating at the tip. The critical stress intensity factor is smaller than that of the low loading rate (Fig. 8).

The influence of the loading rate can be represented in terms of the resistance curves shown in Fig. 10, which present the loading level versus the length of the craze and the crack. Once the craze initiates, the load-bearing ability of the craze allows for some R-curve behavior. At the onset of craze breakdown, crack propagation takes place in an unstable manner, as indicated by the constant load level during crack propagation. This steady-state value is taken as the critical stress intensity factor for the corresponding loading rate. For the set of bulk and craze parameters used here, we observe in Fig. 10 that increasing the loading rate results in a decrease of the critical

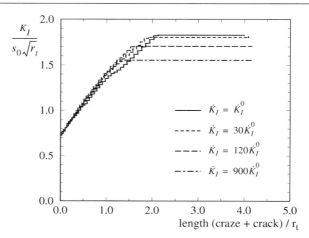

Fig. 10 Influence of the loading rate on the resistance curves using the parameters for set A with $\dot{K}_I^0 \approx 3 \times 10^{-2}$ MPa$\sqrt{\text{m}}$/s and increasing loading rates (from [22])

toughness. In this case, the material shows a ductile-to-brittle transition with increasing loading rate. The results are now considered using the craze parameter set B of Table 3. This set differs from the previous one by the value of A^c which is three times smaller. From the definition of the craze thickening rate (Eq. 22), a smaller A^c value corresponds to a higher thickening rate for a given normal stress. The consequence of a different craze thickening rate is investigated by varying the loading rate within the same range. The resistance curves are shown in Fig. 11. It is observed that the critical stress intensity factor increases with increasing loading rate. In this case, the bulk response is primarily elastic except at the lowest loading rate, in which some plastic deformation emerges at the onset of crack propagation (see Fig. 12a). We notice that shear bands are not originating from the notch tip but away from the crack plane.

For set B, craze thickening is faster and the craze critical thickness is attained at $K_I/(s_0\sqrt{r_t}) \approx 1.32$, which is significantly smaller than the value $K_I/(s_0\sqrt{r_t}) \approx 1.71$ for set A. During crack propagation, some plasticity confined to the craze/crack interface is observed (Fig. 12) but the bulk remains mostly elastic. Therefore, the craze parameters B of Table 3 result in a more brittle response compared to that predicted for the craze parameters A (see Fig. 8b).

The toughening observed in Fig. 11 with increasing loading rate is quite surprising from a standard fracture mechanics point of view on the fracture of viscoplastic materials: increasing the loading rate results in less energy dissipated by plasticity and a more brittle response is expected (if the failure process is assumed to be rate independent). One could also invoke a failure

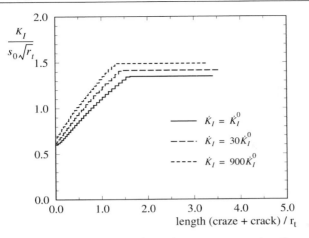

Fig. 11 Influence of the loading rate on the resistance curves using the parameters for set B with $\dot{K}_I^0 \approx 3 \times 10^{-2}$ MPa$\sqrt{\text{m}}$/s and increasing loading rates (from [22])

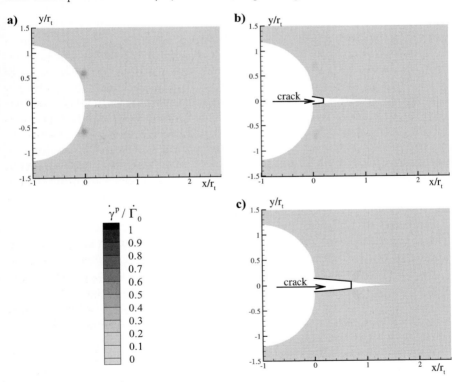

Fig. 12 Instantaneous plastic deformation for the set of craze parameters B at loading rate $\dot{K}_I^0 \approx 3 \times 10^{-2}$ MPa$\sqrt{\text{m}}$/s. **a** Prior to craze fibril breakdown; **b**, **c** during crack propagation, with $K_I/(s_0\sqrt{r_t}) \approx 1.32$ (from [22])

criterion based on a critical crack opening displacement (COD) argument, so that the toughness would increase with loading rate in the case of viscoelastic material. However, in glassy polymers, crazing precedes crack formation which is related to the fibril breakdown. The related parameter Δ^{cr} resembles a COD criterion, but the craze thickening rate (Eq. 22) governs the conditions for crack propagation at Δ^{cr}, independent of the bulk response.

To really understand the effect of the loading rate, we need to distinguish the three timescales introduced by (1) the loading rate $\dot{K}_I$, (2) the reference shear rate $\dot{\gamma}_0$ in the bulk viscoplasticity of Eq. 3, and (3) the $\dot{\Delta}_0$ appearing in Eq. 22 for the crazing time dependence. This set implies two independent ratios. An increase of the loading rate is equivalent to a decrease of the bulk viscoplasticity since the ratio $\dot{K}_I/\dot{\gamma}_0$ increases. Hence, it becomes immediately clear that the material response tends to be less viscoplastic. Furthermore, the ratio $\dot{K}_I/\dot{\Delta}_0$ increases when the loading rate increases, which is equivalent to a reduced craze widening activity for the same applied loading rate. Therefore, craze breakdown is delayed in time, thus leading to a higher K_I^{cr}. There is also the ratio $\dot{\gamma}_0/\dot{\Delta}_0$ which is an indicator of the competition between shear yielding and crazing. If it increases, it means that the kinetics for shear yielding ($\dot{\gamma}_0$) becomes faster than the crazing process and plasticity is promoted.

Therefore, increasing the loading rate results in (1) decreasing plasticity in the bulk and (2) increasing the load level at which craze fibrils break down. The first case is illustrated for the set A of craze parameters for which a ductile-to-brittle transition is observed, because the amount of plasticity prior to crack propagation is diminished as the loading rate increases. For set B, a toughening is observed with increasing loading rate because of the viscoplastic response of the craze thickening process while the bulk is essentially elastic. Both situations have been observed experimentally, with case A resembling the response of PC with blunted notches [1] while case B appears similar to the observations of the toughness increase with loading rate reported in [29, 30] for PMMA.

Estevez et al. [22] also investigated the influence of $\dot{\Delta}_0$ (Eq. 22) and the critical craze thickness Δ^{cr}. Increasing $\dot{\Delta}_0$ corresponds to an increase in the kinetics of craze thickening, so that the condition for craze fibril breakdown is reached earlier and plasticity is prevented, as expected from comparing the timescales. Increasing Δ^{cr} delays the condition for fibril breakdown and crack propagation. The load-bearing ability of the crazes is maintained for a larger craze thickness, which requires a higher load level before craze fibrils break down. Such an increase in Δ^{cr} can also promote plasticity, thus providing an additional increase of the toughness.

In conclusion, the cohesive surface description presented in the foregoing sections appears suitable for capturing a ductile-to-brittle transition with increasing loading rate, and for predicting a toughening effect when the bulk is essentially elastic. These trends are reported experimentally and a calibration of the parameters used in the cohesive zone description is presented in [64].

5.3
Coupled Thermomechanical Analysis

Beyond the classical ductile-to-brittle transition, there is a second but *opposite* transition from brittle to ductile at even higher loading rates. This effect is observed for intermediate loadings between quasistatic and dynamic conditions [59–62]. In this regime, noticeable temperature variations have been recorded prior to and during crack propagation [2, 3, 58, 59]. The temperature variations can originate from heat generated by plastic dissipation during crazing and/or shear yielding. The temperature increase recorded for dynamic crack propagation of PMMA is about hundred(s) of Kelvins [3, 59], so that the temperature can exceed the glass transition temperature T_g. It is worth noting that thermoelastic cooling by about -20 to -30 K prior to crack propagation has been evidenced by Rittel [58].

As a first incursion into the thermomechanical analysis of the problem, we present recent results [57] in which only thermoplastic effects are accounted for. The related temperature variations appear larger than those from thermoelastic effects, and are expected to be of major importance in the competition between shear yielding and crazing. The influence of thermoelastic effects will be briefly discussed at the end of this section.

The coupled problem formulation is presented in Sect. 5.1. The loading rates for which thermal diffusion needs to be considered can be estimated for a one-dimensional problem as in [9, 57]. This leads to the nondimensional quantity κ which compares a characteristic timescale t_0 associated with the loading conditions to the time for heat to diffuse over a characteristic length L_0 as:

$$\kappa = \frac{kt_0}{\rho c_v L_0^2}. \quad (32)$$

For $\kappa \gg 1$, isothermal conditions prevail, while $\kappa \ll 1$ when the situation is adiabatic. The characteristic timescale t_0 for the present study is defined as the time to attain the material toughness K_I^{cr} for a given loading rate, i.e., $t_0 = K_I^{cr}/\dot{K}_I$. The characteristic length L_0 is taken as the size of the plastic zone of a perfectly plastic material with yield stress s_0 so that $L_0 = (K_I/s_0)^2$ [57]. For $\kappa \approx 1$, heat conduction needs to be accounted for and this condition results in the estimation of

$$\dot{K}_I = \frac{ks_0^4}{\rho c_v K_I^3} \approx 100\ \mathrm{MPa}\sqrt{m}/s \quad (33)$$

beyond which a coupled thermomechanical analysis is required. This value is obtained by taking a typical critical toughness of $1\ \mathrm{MPa}\sqrt{m}/s$ and $s_0 = 119.5$ MPa; ρ, c_v, and k are those of Table 2. On this basis, we will consider loading rates between $\dot{K}_I = 300$ and $3000\ \mathrm{MPa}\sqrt{m}/s$, for which crazing initiates at the notch root and the effective yield strength of the material is so high that no significant shear yielding takes place. Higher loading rates would require the account of dynamic effects which therefore will not be considered.

Figure 13 shows the temperature distribution for two different loading rates at the moment that breakdown of fibrils starts. Below these loading rates, the temperature rise is negligible when crack propagation takes place. The temperature distributions in Fig. 13 show that a noticeable variation is located along the faces of the craze, with heat generated during craze thickening diffusing into the material. No temperature rise appears at the craze tip, where further craze initiation occurs. Therefore, even though the craze initiation criterion [24, 25] (Eq. 11) involves temperature-dependent parameters A^0 and B^0, initiation is not affected by heat generated by the crazing process itself.

Figure 14 shows the craze growth resistance curves for the above loading rates together with that for $\dot{K}_I = 30$ MPa$\sqrt{m}$/s from the isothermal analysis in Fig. 11 ($\dot{K}_I = 900\dot{K}_I^0$) [22] for which isothermal conditions prevail. As the loading rate increases, K_I^{cr} remains constant. Toughening caused by temperature effects is not observed, even when the local temperature increases at the highest loading rates.

The temperature distribution during crack propagation is shown in Fig. 15. As the crack advances, the heat continues to diffuse along the normal to the craze surfaces but the size of the hot zone remains comparable to that of the craze thickness. The maximum temperature increase is located at the crack/craze interface, where the craze thickening and related heat flux into the bulk are maxima. At this location, the temperature reaches the glass transition temperature T_g but plasticity is not enhanced in the bulk, which remains primarily elastic during crack propagation.

Experiments by Döll and Könczöl [29, 30] revealed that the critical craze thickness Δ^{cr} is temperature dependent in some cases. To get some feeling for its influence, we consider the case where Δ^{cr} varies linearly from its value

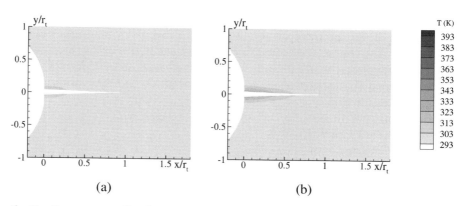

Fig. 13 Temperature distributions at the onset of craze fibril breakdown for **a** $\dot{K}_I = 300$ MPa$\sqrt{m}$/s and **b** $\dot{K}_I = 3000$ MPa$\sqrt{m}$/s (from [57])

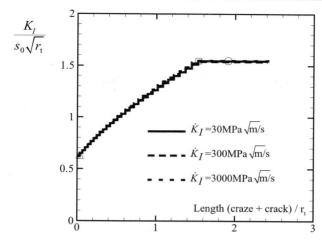

Fig. 14 Craze–crack resistance curves for isothermal conditions with $\dot{K}_I = 30$ MPa$\sqrt{m}$/s and temperature-dependent stress-displacement fields for $\dot{K}_I = 300$ and 3000 MPa$\sqrt{m}$/s. The *square* corresponds to the onset of unstable crack propagation, defining K_I^{cr}

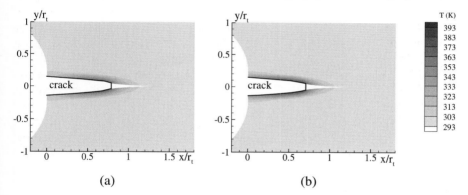

Fig. 15 Temperature distributions during crack propagation for **a** $\dot{K}_I = 300$ MPa$\sqrt{m}$/s and **b** $\dot{K}_I = 3000$ MPa$\sqrt{m}$/s, for a constant craze fibril breakdown

at room temperature to twice that value at T_g. As the loading rate increases, the temperature rise increases as well, and a higher Δ_n^{cr} is observed. This has a direct influence on the resistance curves, yielding a higher K_I^{cr} as $\Delta_n^{cr}(T)$ increases. This is demonstrated in Fig. 16 for $\dot{K}_I = 3000$ MPa$\sqrt{m}$/s, for which K_I^{cr} increases by about 20%.

The temperature distribution at craze breakdown and during crack propagation is shown in Fig. 17. As the crack advances, the temperature reaches the glass transition temperature at the location of the crack–craze transition, where plastic dissipation caused by craze thickening is maximum. However, this remains confined to a small volume around the crack–craze surfaces (see Fig. 17) so that no plasticity on a larger scale is promoted.

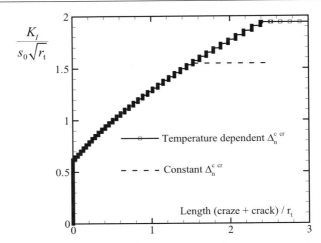

Fig. 16 Craze–crack resistance curves for a constant and a temperature-dependent craze critical thickness

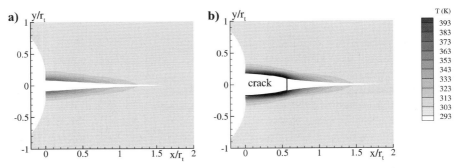

Fig. 17 Temperature distributions **a** at the onset of craze fibril breakdown and **b** during crack propagation for $\dot{K}_I = 3000$ MPa$\sqrt{m}$/s, when a temperature-dependent critical craze thickness is considered

These observations tend to invalidate the interpretation of the toughness increase with loading rate by Williams and Hodgkinson [63]. Their description is based on the idea that crack propagation switches from isothermal to adiabatic conditions when the loading rate increases from static to dynamic conditions. Heat is assumed to originate from crazing, and its diffusion around the craze surfaces is supposed to significantly reduce the yield stress of the bulk material surrounding the craze. Then, thermal effects are assumed to be large enough to promote plasticity in the bulk, which results in blunting of the crack. This picture has been named "thermal blunting".

From the present results, and as long as the crack propagates by a single craze, the calculations indicate that thermal blunting alone cannot be invoked as the explanation of the brittle-to-ductile transition observed at high loading rates. However, a temperature-dependent critical craze thickness may give a tendency in the correct direction.

In the vicinity of the crack tip, the local temperatures show a noticeable increase for loading rates higher than $\dot{K}_I = 300$ MPa$\sqrt{m}$/s. The heat generated by the craze thickening results in a hot zone around the craze surfaces, the extent of which is comparable to that of the craze thickness. In an investigation of thermal effects during crack propagation in PMMA, Fuller et al. [3] estimated the dimension of the hot zone normal to the crack path to be about 1–3 μm. The maximum thickness of a craze in PMMA has been estimated to be about 2–3 μm by Döll and Könczöl [29, 30]. Thus, from two separate experiments, the size of the hot zone normal to the crack path and the critical opening observed for PMMA are of the same order of magnitude. The corresponding quantities resulting from our calculations appear to be at least consistent with these observations. In addition, we observe that craze viscoplasticity is the major heat source during fracture once crazing has initiated, since the bulk remains primarily elastic.

The cohesive surface formulation for crazing implemented within a thermomechanical framework provides insight into the heat generation during crack propagation, and indicates that the temperature-dependent critical craze thickness can result in an increase in toughness, but the marked rise reported in [60, 62] probably has another origin. Here, we believe that dynamic effects need to be considered.

Bjerke and Lambros [59] have recently studied thermal effects generated during dynamic fracture by using a thermally dissipative cohesive zone laid along the crack symmetry plane of a preexisting crack within an elastic bulk. Various cohesive surfaces with rate-independent traction-opening laws were used, thus postulating different physical origins of the failure process. The best agreement with the measured temperature distributions was obtained with a formulation in which the normal stress drops with increasing thickening. However, the cohesive zone in [59] is much wider (about a millimeter) than the craze length reported under quasistatic conditions (hundred(s) of microns [29, 30]). This suggests that the failure under dynamic conditions takes place by a different mechanism than that for quasistatic conditions, as evidenced by the need to use two cohesive zones. Evidently, the mechanisms underlying failure in dynamic fracture need to be clarified.

The cohesive surface considered in the foregoing is based on observations made under quasistatic conditions. In particular, the incubation time for craze initiation is neglected and a critical stress state for craze nucleation is used (Eq. 11). For dynamic loading, a time-dependent craze initiation criterion is to be included in the kinetics, since the characteristic timescale associated with the loading can be comparable to that involved in the craze nucleation process. If the time for craze initiation is accounted for, another timescale is involved in the competition between crazing and shear yielding that determines whether or not crazing takes place. Therefore, a switch from crazing to shear yielding may provide a possible interpretation of the transition from brittle to ductile at high loading rates.

In the present investigation, thermoelastic cooling has not been considered in the heat equation (Eq. 29), since its contribution is expected to be of minor importance when compared to that due to plastic dissipation of crazing or shear yielding. This point may be questionable for materials like PMMA or polystyrene, which exhibit a secondary transition temperature T_β close to room temperature. Schirrer [45] has observed that a transition from single to multiple crazes operates when the temperature decreases below T_β; this also results in an increase in the toughness. In a study devoted to dynamic fracture of PMMA, Rittel [58] has shown that crack tip cooling of about 30 K precedes crack propagation. Therefore, for those glassy polymers having a secondary transition β operating close to room temperature, thermoelastic cooling could promote multiple crazing or inhibit craze initiation to promote shear yielding instead, again resulting in a higher fracture toughness.

6
Conclusion and Future Trends

A description of crazing with a cohesive surface appears appropriate for the crazes observed in glassy polymers, since the trends reported experimentally are quite well captured. The cohesive surface model distinguishes the three steps of crazing (initiation, thickening, and breakdown) and is flexible enough to incorporate more sophisticated formulations of one of these stages when available.

As the craze microstructure is intrinsically discrete rather than continuous, the connection between the variables in the cohesive surface model and molecular characteristics, such as molecular weight, entanglement density or, in more general terms, molecular mobility, is expected to emerge from discrete analyses like the spring network model in [52, 53] or from molecular dynamics as in [49, 50]. Such a connection is currently under development between the critical craze thickness and the characteristics of the fibril structure, and similar developments are expected for the description of the craze kinetics on the basis of molecular dynamics calculations.

From an experimental point of view, the analysis of crazing has focused on the description of the process along the direction normal to the craze surfaces, while the tangential displacement has received little attention. Some ideas have already been reported in studies devoted to craze modeling [20, 21, 55], not least because it is necessary to complete the description. But also, it is necessary to gain more insight into this tangential mode because of its importance when crazing is analyzed in polymer blends or in more complex loadings than the popular mode I when crazing is concerned. How the tangential displacements affect craze thickening and/or breakdown is certainly one of the topics to be addressed in the future. The lateral load-bearing ability of cross-tie fibrils has already been of major importance in the mechanism of fibril breakdown. However, the way in which a lateral displacement affects the breakdown of craze fibrils remains to be clarified.

All these developments are contributing to the evolution of a physically based description of crazing, in which the methodology of cohesive surfaces provides the necessary scale transition from discrete to continuous descriptions.

References

1. Ishikawa M, Narisawa I (1977) J Polym Sci 15:1791
2. Döll W (1973) Eng Fract Mech 5:229
3. Fuller KNG, Fox PG, Field JE (1975) Proc R Soc Lond A 341:537
4. Haward RN, Thackray G (1968) Proc R Soc Lond A 302:453
5. Boyce MC, Parks DM, Argon AS (1988) Mech Mater 7:15
6. Wu PD, Van der Giessen E (1993) J Mech Phys Solids 41:427
7. Wu PD, Van der Giessen E (1996) Eur J Mech 15:799
8. Van der Giessen E (1997) Eur J Mech 16:87
9. Basu S, Van der Giessen E (2002) Int J Plasticity 18:1395
10. Argon AS (1973) Philos Mag 28:839
11. Arruda EM, Boyce MC (1993) J Mech Phys Solids 41:389
12. Raha S, Bowden PB (1972) Polymer 13:174
13. Arruda EM, Boyce MC, Jayachandran R (1995) Mech Mater 19:193
14. Agassant JF, Avenas P, Sergent JPh, Carreau PJ (1991) Polymer processing: principles and modelling. Hanser Gardener, Munich
15. Van Krevelen DW (1990) Properties of polymers, 3rd edn. Elsevier, Amsterdam
16. Kambour RP (1973) J Polym Sci 7:1
17. Kausch HH (1987) Polymer fracture, 2nd edn. Springer, Berlin Heidelberg New York
18. Kausch HH (ed) (1983) Crazing in polymers. Adv Polym Sci 52–53. Springer, Berlin Heidelberg New York
19. Kausch HH (ed) (1990) Crazing in polymers, vol 2. Adv Polym Sci 91–92. Springer, Berlin Heidelberg New York
20. Tijssens MGA, Van der Giessen E, Sluys LJ (2000) Mech Mater 32:19
21. Tijssens MGA, Van der Giessen E, Sluys LJ (2000) Int J Solids Struct 37:7307
22. Estevez R, Tijssens MGA, Van der Giessen E (2000) J Mech Phys Solids 48:2585
23. Argon AS, Hannoosh JG (1977) Philos Mag 36:1195
24. Sternstein SS, Ongchin L (1969) Polymer Prepr 10:1117
25. Sternstein SS, Myers FA (1973) J Macromol Sci Phys B 8:539
26. Oxborough RJ, Bowden PB (1973) Philos Mag 28:547
27. Ishikawa M, Narisawa I (1983) J Mater Sci 2826
28. Gearing BP, Anand L (2004) Int J Solids Struct 41:827
29. Döll W (1983) Adv Polym Sci 52–53:106
30. Döll W, Könczöl L (1990) Adv Polym Sci 91–92:138
31. Kramer EJ (1983) Adv Polym Sci 52–53:1
32. Kramer EJ, Berger LL (1990) Adv Polym Sci 91–92:1
33. Dugdale DS (1960) J Mech Phys Solids 8:100
34. Lauterwasser BD, Kramer EJ (1979) Philos Mag A 39:469

35. Brown HR, Kramer EJ (1981) J Macromol Sci Phys B19:487
36. Van der Giessen E, Lai J (1997) Proceedings of the 10th international conference on deformation, yield and fracture of polymers, Cambridge, 35
37. Tijssens MGA, Van der Giessen E (2002) Polymer 43:831
38. Brown HR, Ward IM (1973) Polymer 14:469
39. Marshall GP, Coutts LH, Williams JG (1974) J Mater Sci 9:1409
40. Williams JG (1984) Fracture mechanics of polymers. Ellis Horwood, New York
41. Döll W, Schinker MG, Könczöl L (1979) Int J Fract 15:R145
42. Pitman GL, Ward IM (1979) Polymer 20:895
43. Morgan GP, Ward IM (1977) Polymer 18:87
44. Weidman GW, Döll W (1978) Int J Fract 14:R189
45. Schirrer R (1990) Adv Polym Sci 91–92:215
46. Brown HR (1991) Macromolecules 24:2752
47. Sih GC, Liebowitz H (1968) In: Liebowitz H (ed) Fracture. Academic, San Diego, p 67
48. Wu S (1990) Polym Eng Sci 30:753
49. Rottler J, Robins MO (2002) Phys Rev Lett 89:1955
50. Rottler J, Robins MO (2003) Phys Rev E 68:118
51. Hui CY, Ruina A, Creton C, Kramer EJ (1992) Macromolecules 25:3948
52. Sha Y, Hui CY, Ruina A, Kramer EJ (1995) Macromolecules 28:2450
53. Sha Y, Hui CY, Kramer EJ (1999) J Mater Sci 34:3695
54. Needleman A (1987) J Appl Mech 54:525
55. Socrate S, Boyce MC, Lazzeri A (2001) Mech Mater 33:155
56. Leevers PS (1995) Int J Fract 73:109
57. Estevez R, Basu S, Van der Giessen E (2005) Int J Fract 132:249
58. Rittel D (1998) Int J Solids Struct 35:2959
59. Bjerke TW, Li Z, Lambros J (2003) J Mech Phys Solids 51:1147
60. Wada H (1992) Eng Fract Mech 41:821
61. Wada H, Seika M, Kennedy TC, Calder CA, Murase K(1996) Eng Fract Mech 54:805
62. Rittel D, Maigre H (1996) Mech Mater 23:229
63. Williams JG, Hodgkinson JM (1981) Proc R Soc Lond A 375:231
64. Saad N, Esteves R, Olagnon C, Séguéla R (2005) Proceedings of the 11th international conference of fracture, 4488

Author Index Volumes 101–188

Author Index Volumes 1–100 see Volume 100

de, Abajo, J. and *de la Campa, J. G.*: Processable Aromatic Polyimides. Vol. 140, pp. 23–60.
Abe, A., Furuya, H., Zhou, Z., Hiejima, T. and *Kobayashi, Y.*: Stepwise Phase Transitions of Chain Molecules: Crystallization/Melting via a Nematic Liquid-Crystalline Phase. Vol. 181, pp. 121–152.
Abetz, V. see Förster, S.: Vol. 166, pp. 173–210.
Adolf, D. B. see Ediger, M. D.: Vol. 116, pp. 73–110.
Aharoni, S. M. and *Edwards, S. F.*: Rigid Polymer Networks. Vol. 118, pp. 1–231.
Albertsson, A.-C. and *Varma, I. K.*: Aliphatic Polyesters: Synthesis, Properties and Applications. Vol. 157, pp. 99–138.
Albertsson, A.-C. see Edlund, U.: Vol. 157, pp. 53–98.
Albertsson, A.-C. see Söderqvist Lindblad, M.: Vol. 157, pp. 139–161.
Albertsson, A.-C. see Stridsberg, K. M.: Vol. 157, pp. 27–51.
Albertsson, A.-C. see Al-Malaika, S.: Vol. 169, pp. 177–199.
Al-Malaika, S.: Perspectives in Stabilisation of Polyolefins. Vol. 169, pp. 121–150.
Altstädt, V.: The Influence of Molecular Variables on Fatigue Resistance in Stress Cracking Environments. Vol. 188, pp. 105–152.
Améduri, B., Boutevin, B. and *Gramain, P.*: Synthesis of Block Copolymers by Radical Polymerization and Telomerization. Vol. 127, pp. 87–142.
Améduri, B. and *Boutevin, B.*: Synthesis and Properties of Fluorinated Telechelic Monodispersed Compounds. Vol. 102, pp. 133–170.
Ameduri, B. see Taguet, A.: Vol. 184, pp. 127–211.
Amselem, S. see Domb, A. J.: Vol. 107, pp. 93–142.
Anantawaraskul, S., Soares, J. B. P. and *Wood-Adams, P. M.*: Fractionation of Semicrystalline Polymers by Crystallization Analysis Fractionation and Temperature Rising Elution Fractionation. Vol. 182, pp. 1–54.
Andrady, A. L.: Wavelenght Sensitivity in Polymer Photodegradation. Vol. 128, pp. 47–94.
Andreis, M. and *Koenig, J. L.*: Application of Nitrogen-15 NMR to Polymers. Vol. 124, pp. 191–238.
Angiolini, L. see Carlini, C.: Vol. 123, pp. 127–214.
Anjum, N. see Gupta, B.: Vol. 162, pp. 37–63.
Anseth, K. S., Newman, S. M. and *Bowman, C. N.*: Polymeric Dental Composites: Properties and Reaction Behavior of Multimethacrylate Dental Restorations. Vol. 122, pp. 177–218.
Antonietti, M. see Cölfen, H.: Vol. 150, pp. 67–187.
Aoki, H. see Ito, S.: Vol. 182, pp. 131–170.
Armitage, B. A. see O'Brien, D. F.: Vol. 126, pp. 53–58.
Arndt, M. see Kaminski, W.: Vol. 127, pp. 143–187.
Arnold, A. and *Holm, C.*: Efficient Methods to Compute Long-Range Interactions for Soft Matter Systems. Vol. 185, pp. 59–109.

Arnold Jr., F. E. and *Arnold, F. E.*: Rigid-Rod Polymers and Molecular Composites. Vol. 117, pp. 257–296.

Arora, M. see Kumar, M. N. V. R.: Vol. 160, pp. 45–118.

Arshady, R.: Polymer Synthesis via Activated Esters: A New Dimension of Creativity in Macromolecular Chemistry. Vol. 111, pp. 1–42.

Auer, S. and *Frenkel, D.*: Numerical Simulation of Crystal Nucleation in Colloids. Vol. 173, pp. 149–208.

Auriemma, F., De Rosa, C. and *Corradini, P.*: Solid Mesophases in Semicrystalline Polymers: Structural Analysis by Diffraction Techniques. Vol. 181, pp. 1–74.

Bahar, I., Erman, B. and *Monnerie, L.*: Effect of Molecular Structure on Local Chain Dynamics: Analytical Approaches and Computational Methods. Vol. 116, pp. 145–206.

Baietto-Dubourg, M. C. see Chateauminois, A.: Vol. 188, pp. 153–193.

Ballauff, M. see Dingenouts, N.: Vol. 144, pp. 1–48.

Ballauff, M. see Holm, C.: Vol. 166, pp. 1–27.

Ballauff, M. see Rühe, J.: Vol. 165, pp. 79–150.

Baltá-Calleja, F. J., González Arche, A., Ezquerra, T. A., Santa Cruz, C., Batallón, F., Frick, B. and *López Cabarcos, E.*: Structure and Properties of Ferroelectric Copolymers of Poly(vinylidene) Fluoride. Vol. 108, pp. 1–48.

Baltussen, J. J. M. see Northolt, M. G.: Vol. 178, pp. 1–108.

Barnes, M. D. see Otaigbe, J. U.: Vol. 154, pp. 1–86.

Barsett, H. see Paulsen, S. B.: Vol. 186, pp. 69–101.

Barshtein, G. R. and *Sabsai, O. Y.*: Compositions with Mineralorganic Fillers. Vol. 101, pp. 1–28.

Barton, J. see Hunkeler, D.: Vol. 112, pp. 115–134.

Baschnagel, J., Binder, K., Doruker, P., Gusev, A. A., Hahn, O., Kremer, K., Mattice, W. L., Müller-Plathe, F., Murat, M., Paul, W., Santos, S., Sutter, U. W. and *Tries, V.*: Bridging the Gap Between Atomistic and Coarse-Grained Models of Polymers: Status and Perspectives. Vol. 152, pp. 41–156.

Bassett, D. C.: On the Role of the Hexagonal Phase in the Crystallization of Polyethylene. Vol. 180, pp. 1–16.

Batallán, F. see Baltá-Calleja, F. J.: Vol. 108, pp. 1–48.

Batog, A. E., Pet'ko, I. P. and *Penczek, P.*: Aliphatic-Cycloaliphatic Epoxy Compounds and Polymers. Vol. 144, pp. 49–114.

Baughman, T. W. and *Wagener, K. B.*: Recent Advances in ADMET Polymerization. Vol. 176, pp. 1–42.

Becker, O. and *Simon, G. P.*: Epoxy Layered Silicate Nanocomposites. Vol. 179, pp. 29–82.

Bell, C. L. and *Peppas, N. A.*: Biomedical Membranes from Hydrogels and Interpolymer Complexes. Vol. 122, pp. 125–176.

Bellon-Maurel, A. see Calmon-Decriaud, A.: Vol. 135, pp. 207–226.

Bennett, D. E. see O'Brien, D. F.: Vol. 126, pp. 53–84.

Berry, G. C.: Static and Dynamic Light Scattering on Moderately Concentraded Solutions: Isotropic Solutions of Flexible and Rodlike Chains and Nematic Solutions of Rodlike Chains. Vol. 114, pp. 233–290.

Bershtein, V. A. and *Ryzhov, V. A.*: Far Infrared Spectroscopy of Polymers. Vol. 114, pp. 43–122.

Bhargava, R., Wang, S.-Q. and *Koenig, J. L*: FTIR Microspectroscopy of Polymeric Systems. Vol. 163, pp. 137–191.

Biesalski, M. see Rühe, J.: Vol. 165, pp. 79–150.

Bigg, D. M.: Thermal Conductivity of Heterophase Polymer Compositions. Vol. 119, pp. 1–30.
Binder, K.: Phase Transitions in Polymer Blends and Block Copolymer Melts: Some Recent Developments. Vol. 112, pp. 115–134.
Binder, K.: Phase Transitions of Polymer Blends and Block Copolymer Melts in Thin Films. Vol. 138, pp. 1–90.
Binder, K. see Baschnagel, J.: Vol. 152, pp. 41–156.
Binder, K., Müller, M., Virnau, P. and *González MacDowell, L.*: Polymer+Solvent Systems: Phase Diagrams, Interface Free Energies, and Nucleation. Vol. 173, pp. 1–104.
Bird, R. B. see Curtiss, C. F.: Vol. 125, pp. 1–102.
Biswas, M. and *Mukherjee, A.*: Synthesis and Evaluation of Metal-Containing Polymers. Vol. 115, pp. 89–124.
Biswas, M. and *Sinha Ray, S.*: Recent Progress in Synthesis and Evaluation of Polymer-Montmorillonite Nanocomposites. Vol. 155, pp. 167–221.
Blankenburg, L. see Klemm, E.: Vol. 177, pp. 53–90.
Blumen, A. see Gurtovenko, A. A.: Vol. 182, pp. 171–282.
Bogdal, D., Penczek, P., Pielichowski, J. and *Prociak, A.*: Microwave Assisted Synthesis, Crosslinking, and Processing of Polymeric Materials. Vol. 163, pp. 193–263.
Bohrisch, J., Eisenbach, C. D., Jaeger, W., Mori, H., Müller, A. H. E., Rehahn, M., Schaller, C., Traser, S. and *Wittmeyer, P.*: New Polyelectrolyte Architectures. Vol. 165, pp. 1–41.
Bolze, J. see Dingenouts, N.: Vol. 144, pp. 1–48.
Bosshard, C.: see Gubler, U.: Vol. 158, pp. 123–190.
Boutevin, B. and *Robin, J. J.*: Synthesis and Properties of Fluorinated Diols. Vol. 102, pp. 105–132.
Boutevin, B. see Améduri, B.: Vol. 102, pp. 133–170.
Boutevin, B. see Améduri, B.: Vol. 127, pp. 87–142.
Boutevin, B. see Guida-Pietrasanta, F.: Vol. 179, pp. 1–27.
Boutevin, B. see Taguet, A.: Vol. 184, pp. 127–211.
Bowman, C. N. see Anseth, K. S.: Vol. 122, pp. 177–218.
Boyd, R. H.: Prediction of Polymer Crystal Structures and Properties. Vol. 116, pp. 1–26.
Bracco, S. see Sozzani, P.: Vol. 181, pp. 153–177.
Briber, R. M. see Hedrick, J. L.: Vol. 141, pp. 1–44.
Bronnikov, S. V., Vettegren, V. I. and *Frenkel, S. Y.*: Kinetics of Deformation and Relaxation in Highly Oriented Polymers. Vol. 125, pp. 103–146.
Brown, H. R. see Creton, C.: Vol. 156, pp. 53–135.
Bruza, K. J. see Kirchhoff, R. A.: Vol. 117, pp. 1–66.
Buchmeiser, M. R.: Regioselective Polymerization of 1-Alkynes and Stereoselective Cyclopolymerization of a, w-Heptadiynes. Vol. 176, pp. 89–119.
Budkowski, A.: Interfacial Phenomena in Thin Polymer Films: Phase Coexistence and Segregation. Vol. 148, pp. 1–112.
Bunz, U. H. F.: Synthesis and Structure of PAEs. Vol. 177, pp. 1–52.
Burban, J. H. see Cussler, E. L.: Vol. 110, pp. 67–80.
Burchard, W.: Solution Properties of Branched Macromolecules. Vol. 143, pp. 113–194.
Butté, A. see Schork, F. J.: Vol. 175, pp. 129–255.

Calmon-Decriaud, A., Bellon-Maurel, V., Silvestre, F.: Standard Methods for Testing the Aerobic Biodegradation of Polymeric Materials. Vol. 135, pp. 207–226.
Cameron, N. R. and *Sherrington, D. C.*: High Internal Phase Emulsions (HIPEs)-Structure, Properties and Use in Polymer Preparation. Vol. 126, pp. 163–214.
de la Campa, J. G. see de Abajo, J.: Vol. 140, pp. 23–60.
Candau, F. see Hunkeler, D.: Vol. 112, pp. 115–134.

Canelas, D. A. and *DeSimone, J. M.*: Polymerizations in Liquid and Supercritical Carbon Dioxide. Vol. 133, pp. 103–140.
Canva, M. and *Stegeman, G. I.*: Quadratic Parametric Interactions in Organic Waveguides. Vol. 158, pp. 87–121.
Capek, I.: Kinetics of the Free-Radical Emulsion Polymerization of Vinyl Chloride. Vol. 120, pp. 135–206.
Capek, I.: Radical Polymerization of Polyoxyethylene Macromonomers in Disperse Systems. Vol. 145, pp. 1–56.
Capek, I. and *Chern, C.-S.*: Radical Polymerization in Direct Mini-Emulsion Systems. Vol. 155, pp. 101–166.
Cappella, B. see Munz, M.: Vol. 164, pp. 87–210.
Carlesso, G. see Prokop, A.: Vol. 160, pp. 119–174.
Carlini, C. and *Angiolini, L.*: Polymers as Free Radical Photoinitiators. Vol. 123, pp. 127–214.
Carter, K. R. see Hedrick, J. L.: Vol. 141, pp. 1–44.
Casas-Vazquez, J. see Jou, D.: Vol. 120, pp. 207–266.
Chan, C.-M. and *Li, L.*: Direct Observation of the Growth of Lamellae and Spherulites by AFM. Vol. 188, pp. 1–41.
Chandrasekhar, V.: Polymer Solid Electrolytes: Synthesis and Structure. Vol. 135, pp. 139–206.
Chang, J. Y. see Han, M. J.: Vol. 153, pp. 1–36.
Chang, T.: Recent Advances in Liquid Chromatography Analysis of Synthetic Polymers. Vol. 163, pp. 1–60.
Charleux, B. and *Faust, R.*: Synthesis of Branched Polymers by Cationic Polymerization. Vol. 142, pp. 1–70.
Chateauminois, A. and *Baietto-Dubourg, M. C.*: Fracture of Glassy Polymers Within Sliding Contacts. Vol. 188, pp. 153–193.
Chen, P. see Jaffe, M.: Vol. 117, pp. 297–328.
Chern, C.-S. see Capek, I.: Vol. 155, pp. 101–166.
Chevolot, Y. see Mathieu, H. J.: Vol. 162, pp. 1–35.
Choe, E.-W. see Jaffe, M.: Vol. 117, pp. 297–328.
Chow, P. Y. and *Gan, L. M.*: Microemulsion Polymerizations and Reactions. Vol. 175, pp. 257–298.
Chow, T. S.: Glassy State Relaxation and Deformation in Polymers. Vol. 103, pp. 149–190.
Chujo, Y. see Uemura, T.: Vol. 167, pp. 81–106.
Chung, S.-J. see Lin, T.-C.: Vol. 161, pp. 157–193.
Chung, T.-S. see Jaffe, M.: Vol. 117, pp. 297–328.
Clarke, N.: Effect of Shear Flow on Polymer Blends. Vol. 183, pp. 127–173.
Cölfen, H. and *Antonietti, M.*: Field-Flow Fractionation Techniques for Polymer and Colloid Analysis. Vol. 150, pp. 67–187.
Colmenero, J. see Richter, D.: Vol. 174, pp. 1–221.
Comanita, B. see Roovers, J.: Vol. 142, pp. 179–228.
Comotti, A. see Sozzani, P.: Vol. 181, pp. 153–177.
Connell, J. W. see Hergenrother, P. M.: Vol. 117, pp. 67–110.
Corradini, P. see Auriemma, F.: Vol. 181, pp. 1–74.
Creton, C., Kramer, E. J., Brown, H. R. and *Hui, C.-Y.*: Adhesion and Fracture of Interfaces Between Immiscible Polymers: From the Molecular to the Continuum Scale. Vol. 156, pp. 53–135.
Criado-Sancho, M. see Jou, D.: Vol. 120, pp. 207–266.
Curro, J. G. see Schweizer, K. S.: Vol. 116, pp. 319–378.

Curtiss, C. F. and *Bird, R. B.*: Statistical Mechanics of Transport Phenomena: Polymeric Liquid Mixtures. Vol. 125, pp. 1–102.
Cussler, E. L., Wang, K. L. and *Burban, J. H.*: Hydrogels as Separation Agents. Vol. 110, pp. 67–80.
Czub, P. see Penczek, P.: Vol. 184, pp. 1–95.

Dalton, L.: Nonlinear Optical Polymeric Materials: From Chromophore Design to Commercial Applications. Vol. 158, pp. 1–86.
Dautzenberg, H. see Holm, C.: Vol. 166, pp. 113–171.
Davidson, J. M. see Prokop, A.: Vol. 160, pp. 119–174.
Den Decker, M. G. see Northolt, M. G.: Vol. 178, pp. 1–108.
Desai, S. M. and *Singh, R. P.*: Surface Modification of Polyethylene. Vol. 169, pp. 231–293.
DeSimone, J. M. see Canelas, D. A.: Vol. 133, pp. 103–140.
DeSimone, J. M. see Kennedy, K. A.: Vol. 175, pp. 329–346.
DiMari, S. see Prokop, A.: Vol. 136, pp. 1–52.
Dimonie, M. V. see Hunkeler, D.: Vol. 112, pp. 115–134.
Dingenouts, N., Bolze, J., Pötschke, D. and *Ballauf, M.*: Analysis of Polymer Latexes by Small-Angle X-Ray Scattering. Vol. 144, pp. 1–48.
Dodd, L. R. and *Theodorou, D. N.*: Atomistic Monte Carlo Simulation and Continuum Mean Field Theory of the Structure and Equation of State Properties of Alkane and Polymer Melts. Vol. 116, pp. 249–282.
Doelker, E.: Cellulose Derivatives. Vol. 107, pp. 199–266.
Dolden, J. G.: Calculation of a Mesogenic Index with Emphasis Upon LC-Polyimides. Vol. 141, pp. 189–245.
Domb, A. J., Amselem, S., Shah, J. and *Maniar, M.*: Polyanhydrides: Synthesis and Characterization. Vol. 107, pp. 93–142.
Domb, A. J. see Kumar, M. N. V. R.: Vol. 160, pp. 45–118.
Doruker, P. see Baschnagel, J.: Vol. 152, pp. 41–156.
Dubois, P. see Mecerreyes, D.: Vol. 147, pp. 1–60.
Dubrovskii, S. A. see Kazanskii, K. S.: Vol. 104, pp. 97–134.
Dudowicz, J. see Freed, K. F.: Vol. 183, pp. 63–126.
Dunkin, I. R. see Steinke, J.: Vol. 123, pp. 81–126.
Dunson, D. L. see McGrath, J. E.: Vol. 140, pp. 61–106.
Dziezok, P. see Rühe, J.: Vol. 165, pp. 79–150.

Eastmond, G. C.: Poly(e-caprolactone) Blends. Vol. 149, pp. 59–223.
Ebringerová, A., Hromádková, Z. and *Heinze, T.*: Hemicellulose. Vol. 186, pp. 1–67.
Economy, J. and *Goranov, K.*: Thermotropic Liquid Crystalline Polymers for High Performance Applications. Vol. 117, pp. 221–256.
Ediger, M. D. and *Adolf, D. B.*: Brownian Dynamics Simulations of Local Polymer Dynamics. Vol. 116, pp. 73–110.
Edlund, U. and *Albertsson, A.-C.*: Degradable Polymer Microspheres for Controlled Drug Delivery. Vol. 157, pp. 53–98.
Edwards, S. F. see Aharoni, S. M.: Vol. 118, pp. 1–231.
Eisenbach, C. D. see Bohrisch, J.: Vol. 165, pp. 1–41.
Endo, T. see Yagci, Y.: Vol. 127, pp. 59–86.
Engelhardt, H. and *Grosche, O.*: Capillary Electrophoresis in Polymer Analysis. Vol. 150, pp. 189–217.
Engelhardt, H. and *Martin, H.*: Characterization of Synthetic Polyelectrolytes by Capillary Electrophoretic Methods. Vol. 165, pp. 211–247.

Eriksson, P. see Jacobson, K.: Vol. 169, pp. 151–176.
Erman, B. see Bahar, I.: Vol. 116, pp. 145–206.
Eschner, M. see Spange, S.: Vol. 165, pp. 43–78.
Estel, K. see Spange, S.: Vol. 165, pp. 43–78.
Estevez, R. and *Van der Giessen, E.*: Modeling and Computational Analysis of Fracture of Glassy Polymers. Vol. 188, pp. 195–234
Ewen, B. and *Richter, D.*: Neutron Spin Echo Investigations on the Segmental Dynamics of Polymers in Melts, Networks and Solutions. Vol. 134, pp. 1–130.
Ezquerra, T. A. see Baltá-Calleja, F. J.: Vol. 108, pp. 1–48.

Fatkullin, N. see Kimmich, R.: Vol. 170, pp. 1–113.
Faust, R. see Charleux, B.: Vol. 142, pp. 1–70.
Faust, R. see Kwon, Y.: Vol. 167, pp. 107–135.
Fekete, E. see Pukánszky, B.: Vol. 139, pp. 109–154.
Fendler, J. H.: Membrane-Mimetic Approach to Advanced Materials. Vol. 113, pp. 1–209.
Fetters, L. J. see Xu, Z.: Vol. 120, pp. 1–50.
Fontenot, K. see Schork, F. J.: Vol. 175, pp. 129–255.
Förster, S., Abetz, V. and *Müller, A. H. E.*: Polyelectrolyte Block Copolymer Micelles. Vol. 166, pp. 173–210.
Förster, S. and *Schmidt, M.*: Polyelectrolytes in Solution. Vol. 120, pp. 51–134.
Freed, K. F. and *Dudowicz, J.*: Influence of Monomer Molecular Structure on the Miscibility of Polymer Blends. Vol. 183, pp. 63–126.
Freire, J. J.: Conformational Properties of Branched Polymers: Theory and Simulations. Vol. 143, pp. 35–112.
Frenkel, S. Y. see Bronnikov, S. V.: Vol. 125, pp. 103–146.
Frick, B. see Baltá-Calleja, F. J.: Vol. 108, pp. 1–48.
Fridman, M. L.: see Terent'eva, J. P.: Vol. 101, pp. 29–64.
Fuchs, G. see Trimmel, G.: Vol. 176, pp. 43–87.
Fukui, K. see Otaigbe, J. U.: Vol. 154, pp. 1–86.
Funke, W.: Microgels-Intramolecularly Crosslinked Macromolecules with a Globular Structure. Vol. 136, pp. 137–232.
Furusho, Y. see Takata, T.: Vol. 171, pp. 1–75.
Furuya, H. see Abe, A.: Vol. 181, pp. 121–152.

Galina, H.: Mean-Field Kinetic Modeling of Polymerization: The Smoluchowski Coagulation Equation. Vol. 137, pp. 135–172.
Gan, L. M. see Chow, P. Y.: Vol. 175, pp. 257–298.
Ganesh, K. see Kishore, K.: Vol. 121, pp. 81–122.
Gaw, K. O. and *Kakimoto, M.*: Polyimide-Epoxy Composites. Vol. 140, pp. 107–136.
Geckeler, K. E. see Rivas, B.: Vol. 102, pp. 171–188.
Geckeler, K. E.: Soluble Polymer Supports for Liquid-Phase Synthesis. Vol. 121, pp. 31–80.
Gedde, U. W. and *Mattozzi, A.*: Polyethylene Morphology. Vol. 169, pp. 29–73.
Gehrke, S. H.: Synthesis, Equilibrium Swelling, Kinetics Permeability and Applications of Environmentally Responsive Gels. Vol. 110, pp. 81–144.
Geil, P. H., Yang, J., Williams, R. A., Petersen, K. L., Long, T.-C. and *Xu, P.*: Effect of Molecular Weight and Melt Time and Temperature on the Morphology of Poly(tetrafluorethylene). Vol. 180, pp. 89–159.
de Gennes, P.-G.: Flexible Polymers in Nanopores. Vol. 138, pp. 91–106.
Georgiou, S.: Laser Cleaning Methodologies of Polymer Substrates. Vol. 168, pp. 1–49.
Geuss, M. see Munz, M.: Vol. 164, pp. 87–210.

Giannelis, E. P., Krishnamoorti, R. and *Manias, E.*: Polymer-Silicate Nanocomposites: Model Systems for Confined Polymers and Polymer Brushes. Vol. 138, pp. 107–148.
Van der Giessen, E. see Estevez, R.: Vol. 188, pp. 195–234
Godovsky, D. Y.: Device Applications of Polymer-Nanocomposites. Vol. 153, pp. 163–205.
Godovsky, D. Y.: Electron Behavior and Magnetic Properties Polymer-Nanocomposites. Vol. 119, pp. 79–122.
González Arche, A. see Baltá-Calleja, F. J.: Vol. 108, pp. 1–48.
Goranov, K. see Economy, J.: Vol. 117, pp. 221–256.
Gramain, P. see Améduri, B.: Vol. 127, pp. 87–142.
Grein, C.: Toughness of Neat, Rubber Modified and Filled β-Nucleated Polypropylene: From Fundamentals to Applications. Vol. 188, pp. 43–104
Grest, G. S.: Normal and Shear Forces Between Polymer Brushes. Vol. 138, pp. 149–184.
Grigorescu, G. and *Kulicke, W.-M.*: Prediction of Viscoelastic Properties and Shear Stability of Polymers in Solution. Vol. 152, p. 1–40.
Gröhn, F. see Rühe, J.: Vol. 165, pp. 79–150.
Grosberg, A. and *Nechaev, S.*: Polymer Topology. Vol. 106, pp. 1–30.
Grosche, O. see Engelhardt, H.: Vol. 150, pp. 189–217.
Grubbs, R., Risse, W. and *Novac, B.*: The Development of Well-defined Catalysts for Ring-Opening Olefin Metathesis. Vol. 102, pp. 47–72.
Gubler, U. and *Bosshard, C.*: Molecular Design for Third-Order Nonlinear Optics. Vol. 158, pp. 123–190.
Guida-Pietrasanta, F. and *Boutevin, B.*: Polysilalkylene or Silarylene Siloxanes Said Hybrid Silicones. Vol. 179, pp. 1–27.
van Gunsteren, W. F. see Gusev, A. A.: Vol. 116, pp. 207–248.
Gupta, B. and *Anjum, N.*: Plasma and Radiation-Induced Graft Modification of Polymers for Biomedical Applications. Vol. 162, pp. 37–63.
Gurtovenko, A. A. and *Blumen, A.*: Generalized Gaussian Structures: Models for Polymer Systems with Complex Topologies. Vol. 182, pp. 171–282.
Gusev, A. A., Müller-Plathe, F., van Gunsteren, W. F. and *Suter, U. W.*: Dynamics of Small Molecules in Bulk Polymers. Vol. 116, pp. 207–248.
Gusev, A. A. see Baschnagel, J.: Vol. 152, pp. 41–156.
Guillot, J. see Hunkeler, D.: Vol. 112, pp. 115–134.
Guyot, A. and *Tauer, K.*: Reactive Surfactants in Emulsion Polymerization. Vol. 111, pp. 43–66.

Hadjichristidis, N., Pispas, S., Pitsikalis, M., Iatrou, H. and *Vlahos, C.*: Asymmetric Star Polymers Synthesis and Properties. Vol. 142, pp. 71–128.
Hadjichristidis, N. see Xu, Z.: Vol. 120, pp. 1–50.
Hadjichristidis, N. see Pitsikalis, M.: Vol. 135, pp. 1–138.
Hahn, O. see Baschnagel, J.: Vol. 152, pp. 41–156.
Hakkarainen, M.: Aliphatic Polyesters: Abiotic and Biotic Degradation and Degradation Products. Vol. 157, pp. 1–26.
Hakkarainen, M. and *Albertsson, A.-C.*: Environmental Degradation of Polyethylene. Vol. 169, pp. 177–199.
Halary, J. L. see Monnerie, L.: Vol. 187, pp. 35–213.
Halary, J. L. see Monnerie, L.: Vol. 187, pp. 215–364.
Hall, H. K. see Penelle, J.: Vol. 102, pp. 73–104.
Hamley, I. W.: Crystallization in Block Copolymers. Vol. 148, pp. 113–138.
Hammouda, B.: SANS from Homogeneous Polymer Mixtures: A Unified Overview. Vol. 106, pp. 87–134.
Han, M. J. and *Chang, J. Y.*: Polynucleotide Analogues. Vol. 153, pp. 1–36.

Harada, A.: Design and Construction of Supramolecular Architectures Consisting of Cyclodextrins and Polymers. Vol. 133, pp. 141–192.
Haralson, M. A. see Prokop, A.: Vol. 136, pp. 1–52.
Harding, S. E.: Analysis of Polysaccharides by Ultracentrifugation. Size, Conformation and Interactions in Solution. Vol. 186, pp. 211–254.
Hasegawa, N. see Usuki, A.: Vol. 179, pp. 135–195.
Hassan, C. M. and *Peppas, N. A.*: Structure and Applications of Poly(vinyl alcohol) Hydrogels Produced by Conventional Crosslinking or by Freezing/Thawing Methods. Vol. 153, pp. 37–65.
Hawker, C. J.: Dentritic and Hyperbranched Macromolecules Precisely Controlled Macromolecular Architectures. Vol. 147, pp. 113–160.
Hawker, C. J. see Hedrick, J. L.: Vol. 141, pp. 1–44.
He, G. S. see Lin, T.-C.: Vol. 161, pp. 157–193.
Hedrick, J. L., Carter, K. R., Labadie, J. W., Miller, R. D., Volksen, W., Hawker, C. J., Yoon, D. Y., Russell, T. P., McGrath, J. E. and *Briber, R. M.*: Nanoporous Polyimides. Vol. 141, pp. 1–44.
Hedrick, J. L., Labadie, J. W., Volksen, W. and *Hilborn, J. G.*: Nanoscopically Engineered Polyimides. Vol. 147, pp. 61–112.
Hedrick, J. L. see Hergenrother, P. M.: Vol. 117, pp. 67–110.
Hedrick, J. L. see Kiefer, J.: Vol. 147, pp. 161–247.
Hedrick, J. L. see McGrath, J. E.: Vol. 140, pp. 61–106.
Heine, D. R., Grest, G. S. and *Curro, J. G.*: Structure of Polymer Melts and Blends: Comparison of Integral Equation theory and Computer Sumulation. Vol. 173, pp. 209–249.
Heinrich, G. and *Klüppel, M.*: Recent Advances in the Theory of Filler Networking in Elastomers. Vol. 160, pp. 1–44.
Heinze, T. see Ebringerová, A.: Vol. 186, pp. 1–67.
Heinze, T. see El Seoud, O. A.: Vol. 186, pp. 103–149.
Heller, J.: Poly (Ortho Esters). Vol. 107, pp. 41–92.
Helm, C. A. see Möhwald, H.: Vol. 165, pp. 151–175.
Hemielec, A. A. see Hunkeler, D.: Vol. 112, pp. 115–134.
Hergenrother, P. M., Connell, J. W., Labadie, J. W. and *Hedrick, J. L.*: Poly(arylene ether)s Containing Heterocyclic Units. Vol. 117, pp. 67–110.
Hernández-Barajas, J. see Wandrey, C.: Vol. 145, pp. 123–182.
Hervet, H. see Léger, L.: Vol. 138, pp. 185–226.
Hiejima, T. see Abe, A.: Vol. 181, pp. 121–152.
Hilborn, J. G. see Hedrick, J. L.: Vol. 147, pp. 61–112.
Hilborn, J. G. see Kiefer, J.: Vol. 147, pp. 161–247.
Hillborg, H. see Vancso, G. J.: Vol. 182, pp. 55–129.
Hiramatsu, N. see Matsushige, M.: Vol. 125, pp. 147–186.
Hirasa, O. see Suzuki, M.: Vol. 110, pp. 241–262.
Hirotsu, S.: Coexistence of Phases and the Nature of First-Order Transition in Poly-N-isopropylacrylamide Gels. Vol. 110, pp. 1–26.
Höcker, H. see Klee, D.: Vol. 149, pp. 1–57.
Holm, C. see Arnold, A.: Vol. 185, pp. 59–109.
Holm, C., Hofmann, T., Joanny, J. F., Kremer, K., Netz, R. R., Reineker, P., Seidel, C., Vilgis, T. A. and *Winkler, R. G.*: Polyelectrolyte Theory. Vol. 166, pp. 67–111.
Holm, C., Rehahn, M., Oppermann, W. and *Ballauff, M.*: Stiff-Chain Polyelectrolytes. Vol. 166, pp. 1–27.
Hornsby, P.: Rheology, Compounding and Processing of Filled Thermoplastics. Vol. 139, pp. 155–216.
Houbenov, N. see Rühe, J.: Vol. 165, pp. 79–150.

Hromádková, Z. see Ebringerová, A.: Vol. 186, pp. 1–67.
Huber, K. see Volk, N.: Vol. 166, pp. 29–65.
Hugenberg, N. see Rühe, J.: Vol. 165, pp. 79–150.
Hui, C.-Y. see Creton, C.: Vol. 156, pp. 53–135.
Hult, A., Johansson, M. and *Malmström, E.*: Hyperbranched Polymers. Vol. 143, pp. 1–34.
Hünenberger, P. H.: Thermostat Algorithms for Molecular-Dynamics Simulations. Vol. 173, pp. 105–147.
Hunkeler, D., Candau, F., Pichot, C., Hemielec, A. E., Xie, T. Y., Barton, J., Vaskova, V., Guillot, J., Dimonie, M. V. and *Reichert, K. H.*: Heterophase Polymerization: A Physical and Kinetic Comparision and Categorization. Vol. 112, pp. 115–134.
Hunkeler, D. see Macko, T.: Vol. 163, pp. 61–136.
Hunkeler, D. see Prokop, A.: Vol. 136, pp. 1–52; 53–74.
Hunkeler, D. see Wandrey, C.: Vol. 145, pp. 123–182.

Iatrou, H. see Hadjichristidis, N.: Vol. 142, pp. 71–128.
Ichikawa, T. see Yoshida, H.: Vol. 105, pp. 3–36.
Ihara, E. see Yasuda, H.: Vol. 133, pp. 53–102.
Ikada, Y. see Uyama, Y.: Vol. 137, pp. 1–40.
Ikehara, T. see Jinnuai, H.: Vol. 170, pp. 115–167.
Ilavsky, M.: Effect on Phase Transition on Swelling and Mechanical Behavior of Synthetic Hydrogels. Vol. 109, pp. 173–206.
Imai, Y.: Rapid Synthesis of Polyimides from Nylon-Salt Monomers. Vol. 140, pp. 1–23.
Inomata, H. see Saito, S.: Vol. 106, pp. 207–232.
Inoue, S. see Sugimoto, H.: Vol. 146, pp. 39–120.
Irie, M.: Stimuli-Responsive Poly(N-isopropylacrylamide), Photo- and Chemical-Induced Phase Transitions. Vol. 110, pp. 49–66.
Ise, N. see Matsuoka, H.: Vol. 114, pp. 187–232.
Ishikawa, T.: Advances in Inorganic Fibers. Vol. 178, pp. 109–144.
Ito, H.: Chemical Amplification Resists for Microlithography. Vol. 172, pp. 37–245.
Ito, K. and *Kawaguchi, S.*: Poly(macronomers), Homo- and Copolymerization. Vol. 142, pp. 129–178.
Ito, K. see Kawaguchi, S.: Vol. 175, pp. 299–328.
Ito, S. and *Aoki, H.*: Nano-Imaging of Polymers by Optical Microscopy. Vol. 182, pp. 131–170.
Ito, Y. see Suginome, M.: Vol. 171, pp. 77–136.
Ivanov, A. E. see Zubov, V. P.: Vol. 104, pp. 135–176.

Jacob, S. and *Kennedy, J.*: Synthesis, Characterization and Properties of OCTA-ARM Polyisobutylene-Based Star Polymers. Vol. 146, pp. 1–38.
Jacobson, K., Eriksson, P., Reitberger, T. and *Stenberg, B.*: Chemiluminescence as a Tool for Polyolefin. Vol. 169, pp. 151–176.
Jaeger, W. see Bohrisch, J.: Vol. 165, pp. 1–41.
Jaffe, M., Chen, P., Choe, E.-W., Chung, T.-S. and *Makhija, S.*: High Performance Polymer Blends. Vol. 117, pp. 297–328.
Jancar, J.: Structure-Property Relationships in Thermoplastic Matrices. Vol. 139, pp. 1–66.
Jen, A. K.-Y. see Kajzar, F.: Vol. 161, pp. 1–85.
Jerome, R. see Mecerreyes, D.: Vol. 147, pp. 1–60.
de Jeu, W. H. see Li, L.: Vol. 181, pp. 75–120.
Jiang, M., Li, M., Xiang, M. and *Zhou, H.*: Interpolymer Complexation and Miscibility and Enhancement by Hydrogen Bonding. Vol. 146, pp. 121–194.
Jin, J. see Shim, H.-K.: Vol. 158, pp. 191–241.

Jinnai, H., Nishikawa, Y., Ikehara, T. and *Nishi, T.*: Emerging Technologies for the 3D Analysis of Polymer Structures. Vol. 170, pp. 115–167.

Jo, W. H. and *Yang, J. S.*: Molecular Simulation Approaches for Multiphase Polymer Systems. Vol. 156, pp. 1–52.

Joanny, J.-F. see Holm, C.: Vol. 166, pp. 67–111.

Joanny, J.-F. see Thünemann, A. F.: Vol. 166, pp. 113–171.

Johannsmann, D. see Rühe, J.: Vol. 165, pp. 79–150.

Johansson, M. see Hult, A.: Vol. 143, pp. 1–34.

Joos-Müller, B. see Funke, W.: Vol. 136, pp. 137–232.

Jou, D., Casas-Vazquez, J. and *Criado-Sancho, M.*: Thermodynamics of Polymer Solutions under Flow: Phase Separation and Polymer Degradation. Vol. 120, pp. 207–266.

Kaetsu, I.: Radiation Synthesis of Polymeric Materials for Biomedical and Biochemical Applications. Vol. 105, pp. 81–98.

Kaji, K. see Kanaya, T.: Vol. 154, pp. 87–141.

Kajzar, F., Lee, K.-S. and *Jen, A. K.-Y.*: Polymeric Materials and their Orientation Techniques for Second-Order Nonlinear Optics. Vol. 161, pp. 1–85.

Kakimoto, M. see Gaw, K. O.: Vol. 140, pp. 107–136.

Kaminski, W. and *Arndt, M.*: Metallocenes for Polymer Catalysis. Vol. 127, pp. 143–187.

Kammer, H. W., Kressler, H. and *Kummerloewe, C.*: Phase Behavior of Polymer Blends – Effects of Thermodynamics and Rheology. Vol. 106, pp. 31–86.

Kanaya, T. and *Kaji, K.*: Dynamcis in the Glassy State and Near the Glass Transition of Amorphous Polymers as Studied by Neutron Scattering. Vol. 154, pp. 87–141.

Kandyrin, L. B. and *Kuleznev, V. N.*: The Dependence of Viscosity on the Composition of Concentrated Dispersions and the Free Volume Concept of Disperse Systems. Vol. 103, pp. 103–148.

Kaneko, M. see Ramaraj, R.: Vol. 123, pp. 215–242.

Kang, E. T., Neoh, K. G. and *Tan, K. L.*: X-Ray Photoelectron Spectroscopic Studies of Electroactive Polymers. Vol. 106, pp. 135–190.

Karlsson, S. see Söderqvist Lindblad, M.: Vol. 157, pp. 139–161.

Karlsson, S.: Recycled Polyolefins. Material Properties and Means for Quality Determination. Vol. 169, pp. 201–229.

Kato, K. see Uyama, Y.: Vol. 137, pp. 1–40.

Kato, M. see Usuki, A.: Vol. 179, pp. 135–195.

Kausch, H.-H. and *Michler, G. H.*: The Effect of Time on Crazing and Fracture. Vol. 187, pp. 1–33.

Kausch, H.-H. see Monnerie, L. Vol. 187, pp. 215–364.

Kautek, W. see Krüger, J.: Vol. 168, pp. 247–290.

Kawaguchi, S. see Ito, K.: Vol. 142, pp. 129–178.

Kawaguchi, S. and *Ito, K.*: Dispersion Polymerization. Vol. 175, pp. 299–328.

Kawata, S. see Sun, H.-B.: Vol. 170, pp. 169–273.

Kazanskii, K. S. and *Dubrovskii, S. A.*: Chemistry and Physics of Agricultural Hydrogels. Vol. 104, pp. 97–134.

Kennedy, J. P. see Jacob, S.: Vol. 146, pp. 1–38.

Kennedy, J. P. see Majoros, I.: Vol. 112, pp. 1–113.

Kennedy, K. A., Roberts, G. W. and *DeSimone, J. M.*: Heterogeneous Polymerization of Fluoroolefins in Supercritical Carbon Dioxide. Vol. 175, pp. 329–346.

Khokhlov, A., Starodybtzev, S. and *Vasilevskaya, V.*: Conformational Transitions of Polymer Gels: Theory and Experiment. Vol. 109, pp. 121–172.

Kiefer, J., Hedrick, J. L. and *Hiborn, J. G.*: Macroporous Thermosets by Chemically Induced Phase Separation. Vol. 147, pp. 161–247.
Kihara, N. see Takata, T.: Vol. 171, pp. 1–75.
Kilian, H. G. and *Pieper, T.*: Packing of Chain Segments. A Method for Describing X-Ray Patterns of Crystalline, Liquid Crystalline and Non-Crystalline Polymers. Vol. 108, pp. 49–90.
Kim, J. see Quirk, R. P.: Vol. 153, pp. 67–162.
Kim, K.-S. see Lin, T.-C.: Vol. 161, pp. 157–193.
Kimmich, R. and *Fatkullin, N.*: Polymer Chain Dynamics and NMR. Vol. 170, pp. 1–113.
Kippelen, B. and *Peyghambarian, N.*: Photorefractive Polymers and their Applications. Vol. 161, pp. 87–156.
Kirchhoff, R. A. and *Bruza, K. J.*: Polymers from Benzocyclobutenes. Vol. 117, pp. 1–66.
Kishore, K. and *Ganesh, K.*: Polymers Containing Disulfide, Tetrasulfide, Diselenide and Ditelluride Linkages in the Main Chain. Vol. 121, pp. 81–122.
Kitamaru, R.: Phase Structure of Polyethylene and Other Crystalline Polymers by Solid-State 13C/MNR. Vol. 137, pp. 41–102.
Klapper, M. see Rusanov, A. L.: Vol. 179, pp. 83–134.
Klee, D. and *Höcker, H.*: Polymers for Biomedical Applications: Improvement of the Interface Compatibility. Vol. 149, pp. 1–57.
Klemm, E., Pautzsch, T. and *Blankenburg, L.*: Organometallic PAEs. Vol. 177, pp. 53–90.
Klier, J. see Scranton, A. B.: Vol. 122, pp. 1–54.
v. Klitzing, R. and *Tieke, B.*: Polyelectrolyte Membranes. Vol. 165, pp. 177–210.
Klüppel, M.: The Role of Disorder in Filler Reinforcement of Elastomers on Various Length Scales. Vol. 164, pp. 1–86.
Klüppel, M. see Heinrich, G.: Vol. 160, pp. 1–44.
Knuuttila, H., Lehtinen, A. and *Nummila-Pakarinen, A.*: Advanced Polyethylene Technologies – Controlled Material Properties. Vol. 169, pp. 13–27.
Kobayashi, S., Shoda, S. and *Uyama, H.*: Enzymatic Polymerization and Oligomerization. Vol. 121, pp. 1–30.
Kobayashi, T. see Abe, A.: Vol. 181, pp. 121–152.
Köhler, W. and *Schäfer, R.*: Polymer Analysis by Thermal-Diffusion Forced Rayleigh Scattering. Vol. 151, pp. 1–59.
Koenig, J. L. see Bhargava, R.: Vol. 163, pp. 137–191.
Koenig, J. L. see Andreis, M.: Vol. 124, pp. 191–238.
Koike, T.: Viscoelastic Behavior of Epoxy Resins Before Crosslinking. Vol. 148, pp. 139–188.
Kokko, E. see Löfgren, B.: Vol. 169, pp. 1–12.
Kokufuta, E.: Novel Applications for Stimulus-Sensitive Polymer Gels in the Preparation of Functional Immobilized Biocatalysts. Vol. 110, pp. 157–178.
Konno, M. see Saito, S.: Vol. 109, pp. 207–232.
Konradi, R. see Rühe, J.: Vol. 165, pp. 79–150.
Kopecek, J. see Putnam, D.: Vol. 122, pp. 55–124.
Koßmehl, G. see Schopf, G.: Vol. 129, pp. 1–145.
Kostoglodov, P. V. see Rusanov, A. L.: Vol. 179, pp. 83–134.
Kozlov, E. see Prokop, A.: Vol. 160, pp. 119–174.
Kramer, E. J. see Creton, C.: Vol. 156, pp. 53–135.
Kremer, K. see Baschnagel, J.: Vol. 152, pp. 41–156.
Kremer, K. see Holm, C.: Vol. 166, pp. 67–111.
Kressler, J. see Kammer, H. W.: Vol. 106, pp. 31–86.
Kricheldorf, H. R.: Liquid-Cristalline Polyimides. Vol. 141, pp. 83–188.
Krishnamoorti, R. see Giannelis, E. P.: Vol. 138, pp. 107–148.

Krüger, J. and *Kautek, W.*: Ultrashort Pulse Laser Interaction with Dielectrics and Polymers, Vol. 168, pp. 247–290.
Kuchanov, S. I.: Modern Aspects of Quantitative Theory of Free-Radical Copolymerization. Vol. 103, pp. 1–102.
Kuchanov, S. I.: Principles of Quantitive Description of Chemical Structure of Synthetic Polymers. Vol. 152, pp. 157–202.
Kudaibergennow, S. E.: Recent Advances in Studying of Synthetic Polyampholytes in Solutions. Vol. 144, pp. 115–198.
Kuleznev, V. N. see Kandyrin, L. B.: Vol. 103, pp. 103–148.
Kulichkhin, S. G. see Malkin, A. Y.: Vol. 101, pp. 217–258.
Kulicke, W.-M. see Grigorescu, G.: Vol. 152, pp. 1–40.
Kumar, M. N. V. R., Kumar, N., Domb, A. J. and *Arora, M.*: Pharmaceutical Polymeric Controlled Drug Delivery Systems. Vol. 160, pp. 45–118.
Kumar, N. see Kumar, M. N. V. R.: Vol. 160, pp. 45–118.
Kummerloewe, C. see Kammer, H. W.: Vol. 106, pp. 31–86.
Kuznetsova, N. P. see Samsonov, G. V.: Vol. 104, pp. 1–50.
Kwon, Y. and *Faust, R.*: Synthesis of Polyisobutylene-Based Block Copolymers with Precisely Controlled Architecture by Living Cationic Polymerization. Vol. 167, pp. 107–135.

Labadie, J. W. see Hergenrother, P. M.: Vol. 117, pp. 67–110.
Labadie, J. W. see Hedrick, J. L.: Vol. 141, pp. 1–44.
Labadie, J. W. see Hedrick, J. L.: Vol. 147, pp. 61–112.
Lamparski, H. G. see O'Brien, D. F.: Vol. 126, pp. 53–84.
Laschewsky, A.: Molecular Concepts, Self-Organisation and Properties of Polysoaps. Vol. 124, pp. 1–86.
Laso, M. see Leontidis, E.: Vol. 116, pp. 283–318.
Lauprêtre, F. see Monnerie, L.: Vol. 187, pp. 35–213.
Lazár, M. and *Rychl, R.*: Oxidation of Hydrocarbon Polymers. Vol. 102, pp. 189–222.
Lechowicz, J. see Galina, H.: Vol. 137, pp. 135–172.
Léger, L., Raphaël, E. and *Hervet, H.*: Surface-Anchored Polymer Chains: Their Role in Adhesion and Friction. Vol. 138, pp. 185–226.
Lenz, R. W.: Biodegradable Polymers. Vol. 107, pp. 1–40.
Leontidis, E., de Pablo, J. J., Laso, M. and *Suter, U. W.*: A Critical Evaluation of Novel Algorithms for the Off-Lattice Monte Carlo Simulation of Condensed Polymer Phases. Vol. 116, pp. 283–318.
Lee, B. see Quirk, R. P.: Vol. 153, pp. 67–162.
Lee, K.-S. see Kajzar, F.: Vol. 161, pp. 1–85.
Lee, Y. see Quirk, R. P.: Vol. 153, pp. 67–162.
Lehtinen, A. see Knuuttila, H.: Vol. 169, pp. 13–27.
Leónard, D. see Mathieu, H. J.: Vol. 162, pp. 1–35.
Lesec, J. see Viovy, J.-L.: Vol. 114, pp. 1–42.
Levesque, D. see Weis, J.-J.: Vol. 185, pp. 163–225.
Li, L. and *de Jeu, W. H.*: Flow-induced mesophases in crystallizable polymers. Vol. 181, pp. 75–120.
Li, L. see Chan, C.-M.: Vol. 188, pp. 1–41
Li, M. see Jiang, M.: Vol. 146, pp. 121–194.
Liang, G. L. see Sumpter, B. G.: Vol. 116, pp. 27–72.
Lienert, K.-W.: Poly(ester-imide)s for Industrial Use. Vol. 141, pp. 45–82.
Likhatchev, D. see Rusanov, A. L.: Vol. 179, pp. 83–134.

Lin, J. and *Sherrington, D. C.*: Recent Developments in the Synthesis, Thermostability and Liquid Crystal Properties of Aromatic Polyamides. Vol. 111, pp. 177–220.
Lin, T.-C., Chung, S.-J., Kim, K.-S., Wang, X., He, G. S., Swiatkiewicz, J., Pudavar, H. E. and *Prasad, P. N.*: Organics and Polymers with High Two-Photon Activities and their Applications. Vol. 161, pp. 157–193.
Linse, P.: Simulation of Charged Colloids in Solution. Vol. 185, pp. 111–162.
Lippert, T.: Laser Application of Polymers. Vol. 168, pp. 51–246.
Liu, Y. see Söderqvist Lindblad, M.: Vol. 157, pp. 139–161.
Long, T.-C. see Geil, P. H.: Vol. 180, pp. 89–159.
López Cabarcos, E. see Baltá-Calleja, F. J.: Vol. 108, pp. 1–48.
Lotz, B.: Analysis and Observation of Polymer Crystal Structures at the Individual Stem Level. Vol. 180, pp. 17–44.
Löfgren, B., Kokko, E. and *Seppälä, J.*: Specific Structures Enabled by Metallocene Catalysis in Polyethenes. Vol. 169, pp. 1–12.
Löwen, H. see Thünemann, A. F.: Vol. 166, pp. 113–171.
Luo, Y. see Schork, F. J.: Vol. 175, pp. 129–255.

Macko, T. and *Hunkeler, D.*: Liquid Chromatography under Critical and Limiting Conditions: A Survey of Experimental Systems for Synthetic Polymers. Vol. 163, pp. 61–136.
Majoros, I., Nagy, A. and *Kennedy, J. P.*: Conventional and Living Carbocationic Polymerizations United. I. A Comprehensive Model and New Diagnostic Method to Probe the Mechanism of Homopolymerizations. Vol. 112, pp. 1–113.
Makhija, S. see Jaffe, M.: Vol. 117, pp. 297–328.
Malmström, E. see Hult, A.: Vol. 143, pp. 1–34.
Malkin, A. Y. and *Kulichkhin, S. G.*: Rheokinetics of Curing. Vol. 101, pp. 217–258.
Maniar, M. see Domb, A. J.: Vol. 107, pp. 93–142.
Manias, E. see Giannelis, E. P.: Vol. 138, pp. 107–148.
Martin, H. see Engelhardt, H.: Vol. 165, pp. 211–247.
Marty, J. D. and *Mauzac, M.*: Molecular Imprinting: State of the Art and Perspectives. Vol. 172, pp. 1–35.
Mashima, K., Nakayama, Y. and *Nakamura, A.*: Recent Trends in Polymerization of a-Olefins Catalyzed by Organometallic Complexes of Early Transition Metals. Vol. 133, pp. 1–52.
Mathew, D. see Reghunadhan Nair, C. P.: Vol. 155, pp. 1–99.
Mathieu, H. J., Chevolot, Y, Ruiz-Taylor, L. and *Leónard, D.*: Engineering and Characterization of Polymer Surfaces for Biomedical Applications. Vol. 162, pp. 1–35.
Matsumoto, A.: Free-Radical Crosslinking Polymerization and Copolymerization of Multivinyl Compounds. Vol. 123, pp. 41–80.
Matsumoto, A. see Otsu, T.: Vol. 136, pp. 75–138.
Matsuoka, H. and *Ise, N.*: Small-Angle and Ultra-Small Angle Scattering Study of the Ordered Structure in Polyelectrolyte Solutions and Colloidal Dispersions. Vol. 114, pp. 187–232.
Matsushige, K., Hiramatsu, N. and *Okabe, H.*: Ultrasonic Spectroscopy for Polymeric Materials. Vol. 125, pp. 147–186.
Mattice, W. L. see Rehahn, M.: Vol. 131/132, pp. 1–475.
Mattice, W. L. see Baschnagel, J.: Vol. 152, pp. 41–156.
Mattozzi, A. see Gedde, U. W.: Vol. 169, pp. 29–73.
Mauzac, M. see Marty, J. D.: Vol. 172, pp. 1–35.
Mays, W. see Xu, Z.: Vol. 120, pp. 1–50.
Mays, J. W. see Pitsikalis, M.: Vol. 135, pp. 1–138.
McGrath, J. E. see Hedrick, J. L.: Vol. 141, pp. 1–44.

McGrath, J. E., Dunson, D. L. and *Hedrick, J. L.*: Synthesis and Characterization of Segmented Polyimide-Polyorganosiloxane Copolymers. Vol. 140, pp. 61–106.
McLeish, T. C. B. and *Milner, S. T.*: Entangled Dynamics and Melt Flow of Branched Polymers. Vol. 143, pp. 195–256.
Mecerreyes, D., Dubois, P. and *Jerome, R.*: Novel Macromolecular Architectures Based on Aliphatic Polyesters: Relevance of the Coordination-Insertion Ring-Opening Polymerization. Vol. 147, pp. 1–60.
Mecham, S. J. see McGrath, J. E.: Vol. 140, pp. 61–106.
Menzel, H. see Möhwald, H.: Vol. 165, pp. 151–175.
Meyer, T. see Spange, S.: Vol. 165, pp. 43–78.
Michler, G. H. see Kausch, H.-H.: Vol. 187, pp. 1–33.
Mikos, A. G. see Thomson, R. C.: Vol. 122, pp. 245–274.
Milner, S. T. see McLeish, T. C. B.: Vol. 143, pp. 195–256.
Mison, P. and *Sillion, B.*: Thermosetting Oligomers Containing Maleimides and Nadiimides End-Groups. Vol. 140, pp. 137–180.
Miyasaka, K.: PVA-Iodine Complexes: Formation, Structure and Properties. Vol. 108, pp. 91–130.
Miller, R. D. see Hedrick, J. L.: Vol. 141, pp. 1–44.
Minko, S. see Rühe, J.: Vol. 165, pp. 79–150.
Möhwald, H., Menzel, H., Helm, C. A. and *Stamm, M.*: Lipid and Polyampholyte Monolayers to Study Polyelectrolyte Interactions and Structure at Interfaces. Vol. 165, pp. 151–175.
Monkenbusch, M. see Richter, D.: Vol. 174, pp. 1–221.
Monnerie, L., Halary, J. L. and *Kausch, H.-H.*: Deformation, Yield and Fracture of Amorphous Polymers: Relation to the Secondary Transitions. Vol. 187, pp. 215–364.
Monnerie, L., Lauprêtre, F. and *Halary, J. L.*: Investigation of Solid-State Transitions in Linear and Crosslinked Amorphous Polymers. Vol. 187, pp. 35–213.
Monnerie, L. see Bahar, I.: Vol. 116, pp. 145–206.
Moore, J. S. see Ray, C. R.: Vol. 177, pp. 99–149.
Mori, H. see Bohrisch, J.: Vol. 165, pp. 1–41.
Morishima, Y.: Photoinduced Electron Transfer in Amphiphilic Polyelectrolyte Systems. Vol. 104, pp. 51–96.
Morton, M. see Quirk, R. P.: Vol. 153, pp. 67–162.
Motornov, M. see Rühe, J.: Vol. 165, pp. 79–150.
Mours, M. see Winter, H. H.: Vol. 134, pp. 165–234.
Müllen, K. see Scherf, U.: Vol. 123, pp. 1–40.
Müller, A. H. E. see Bohrisch, J.: Vol. 165, pp. 1–41.
Müller, A. H. E. see Förster, S.: Vol. 166, pp. 173–210.
Müller, M. and *Schmid, F.*: Incorporating Fluctuations and Dynamics in Self-Consistent Field Theories for Polymer Blends. Vol. 185, pp. 1–58.
Müller, M. see Thünemann, A. F.: Vol. 166, pp. 113–171.
Müller-Plathe, F. see Gusev, A. A.: Vol. 116, pp. 207–248.
Müller-Plathe, F. see Baschnagel, J.: Vol. 152, p. 41–156.
Mukerherjee, A. see Biswas, M.: Vol. 115, pp. 89–124.
Munz, M., Cappella, B., Sturm, H., Geuss, M. and *Schulz, E.*: Materials Contrasts and Nanolithography Techniques in Scanning Force Microscopy (SFM) and their Application to Polymers and Polymer Composites. Vol. 164, pp. 87–210.
Murat, M. see Baschnagel, J.: Vol. 152, p. 41–156.
Muzzarelli, C. see Muzzarelli, R. A. A.: Vol. 186, pp. 151–209.

Muzzarelli, R. A. A. and *Muzzarelli, C.*: Chitosan Chemistry: Relevance to the Biomedical Sciences. Vol. 186, pp. 151–209.
Mylnikov, V.: Photoconducting Polymers. Vol. 115, pp. 1–88.

Nagy, A. see Majoros, I.: Vol. 112, pp. 1–11.
Naka, K. see Uemura, T.: Vol. 167, pp. 81–106.
Nakamura, A. see Mashima, K.: Vol. 133, pp. 1–52.
Nakayama, Y. see Mashima, K.: Vol. 133, pp. 1–52.
Narasinham, B. and *Peppas, N. A.*: The Physics of Polymer Dissolution: Modeling Approaches and Experimental Behavior. Vol. 128, pp. 157–208.
Nechaev, S. see Grosberg, A.: Vol. 106, pp. 1–30.
Neoh, K. G. see Kang, E. T.: Vol. 106, pp. 135–190.
Netz, R. R. see Holm, C.: Vol. 166, pp. 67–111.
Netz, R. R. see Rühe, J.: Vol. 165, pp. 79–150.
Newman, S. M. see Anseth, K. S.: Vol. 122, pp. 177–218.
Nijenhuis, K. te: Thermoreversible Networks. Vol. 130, pp. 1–252.
Ninan, K. N. see Reghunadhan Nair, C. P.: Vol. 155, pp. 1–99.
Nishi, T. see Jinnai, H.: Vol. 170, pp. 115–167.
Nishikawa, Y. see Jinnai, H.: Vol. 170, pp. 115–167.
Noid, D. W. see Otaigbe, J. U.: Vol. 154, pp. 1–86.
Noid, D. W. see Sumpter, B. G.: Vol. 116, pp. 27–72.
Nomura, M., Tobita, H. and *Suzuki, K.*: Emulsion Polymerization: Kinetic and Mechanistic Aspects. Vol. 175, pp. 1–128.
Northolt, M. G., Picken, S. J., Den Decker, M. G., Baltussen, J. J. M. and *Schlatmann, R.*: The Tensile Strength of Polymer Fibres. Vol. 178, pp. 1–108.
Novac, B. see Grubbs, R.: Vol. 102, pp. 47–72.
Novikov, V. V. see Privalko, V. P.: Vol. 119, pp. 31–78.
Nummila-Pakarinen, A. see Knuuttila, H.: Vol. 169, pp. 13–27.

O'Brien, D. F., Armitage, B. A., Bennett, D. E. and *Lamparski, H. G.*: Polymerization and Domain Formation in Lipid Assemblies. Vol. 126, pp. 53–84.
Ogasawara, M.: Application of Pulse Radiolysis to the Study of Polymers and Polymerizations. Vol.105, pp. 37–80.
Okabe, H. see Matsushige, K.: Vol. 125, pp. 147–186.
Okada, M.: Ring-Opening Polymerization of Bicyclic and Spiro Compounds. Reactivities and Polymerization Mechanisms. Vol. 102, pp. 1–46.
Okano, T.: Molecular Design of Temperature-Responsive Polymers as Intelligent Materials. Vol. 110, pp. 179–198.
Okay, O. see Funke, W.: Vol. 136, pp. 137–232.
Onuki, A.: Theory of Phase Transition in Polymer Gels. Vol. 109, pp. 63–120.
Oppermann, W. see Holm, C.: Vol. 166, pp. 1–27.
Oppermann, W. see Volk, N.: Vol. 166, pp. 29–65.
Osad'ko, I. S.: Selective Spectroscopy of Chromophore Doped Polymers and Glasses. Vol. 114, pp. 123–186.
Osakada, K. and *Takeuchi, D.*: Coordination Polymerization of Dienes, Allenes, and Methylenecycloalkanes. Vol. 171, pp. 137–194.
Otaigbe, J. U., Barnes, M. D., Fukui, K., Sumpter, B. G. and *Noid, D. W.*: Generation, Characterization, and Modeling of Polymer Micro- and Nano-Particles. Vol. 154, pp. 1–86.
Otsu, T. and *Matsumoto, A.*: Controlled Synthesis of Polymers Using the Iniferter Technique: Developments in Living Radical Polymerization. Vol. 136, pp. 75–138.

de Pablo, J. J. see Leontidis, E.: Vol. 116, pp. 283–318.
Padias, A. B. see Penelle, J.: Vol. 102, pp. 73–104.
Pascault, J.-P. see Williams, R. J. J.: Vol. 128, pp. 95–156.
Pasch, H.: Analysis of Complex Polymers by Interaction Chromatography. Vol. 128, pp. 1–46.
Pasch, H.: Hyphenated Techniques in Liquid Chromatography of Polymers. Vol. 150, pp. 1–66.
Paul, W. see Baschnagel, J.: Vol. 152, pp. 41–156.
Paulsen, S. B. and *Barsett, H.*: Bioactive Pectic Polysaccharides. Vol. 186, pp. 69–101.
Pautzsch, T. see Klemm, E.: Vol. 177, pp. 53–90.
Penczek, P., Czub, P. and *Pielichowski, J.*: Unsaturated Polyester Resins: Chemistry and Technology. Vol. 184, pp. 1–95.
Penczek, P. see Batog, A. E.: Vol. 144, pp. 49–114.
Penczek, P. see Bogdal, D.: Vol. 163, pp. 193–263.
Penelle, J., Hall, H. K., Padias, A. B. and *Tanaka, H.*: Captodative Olefins in Polymer Chemistry. Vol. 102, pp. 73–104.
Peppas, N. A. see Bell, C. L.: Vol. 122, pp. 125–176.
Peppas, N. A. see Hassan, C. M.: Vol. 153, pp. 37–65.
Peppas, N. A. see Narasimhan, B.: Vol. 128, pp. 157–208.
Petersen, K. L. see Geil, P. H.: Vol. 180, pp. 89–159.
Pet'ko, I. P. see Batog, A. E.: Vol. 144, pp. 49–114.
Pheyghambarian, N. see Kippelen, B.: Vol. 161, pp. 87–156.
Pichot, C. see Hunkeler, D.: Vol. 112, pp. 115–134.
Picken, S. J. see Northolt, M. G.: Vol. 178, pp. 1–108.
Pielichowski, J. see Bogdal, D.: Vol. 163, pp. 193–263.
Pielichowski, J. see Penczek, P.: Vol. 184, pp. 1–95.
Pieper, T. see Kilian, H. G.: Vol. 108, pp. 49–90.
Pispas, S. see Pitsikalis, M.: Vol. 135, pp. 1–138.
Pispas, S. see Hadjichristidis, N.: Vol. 142, pp. 71–128.
Pitsikalis, M., Pispas, S., Mays, J. W. and *Hadjichristidis, N.*: Nonlinear Block Copolymer Architectures. Vol. 135, pp. 1–138.
Pitsikalis, M. see Hadjichristidis, N.: Vol. 142, pp. 71–128.
Pleul, D. see Spange, S.: Vol. 165, pp. 43–78.
Plummer, C. J. G.: Microdeformation and Fracture in Bulk Polyolefins. Vol. 169, pp. 75–119.
Pötschke, D. see Dingenouts, N.: Vol. 144, pp. 1–48.
Pokrovskii, V. N.: The Mesoscopic Theory of the Slow Relaxation of Linear Macromolecules. Vol. 154, pp. 143–219.
Pospíšil, J.: Functionalized Oligomers and Polymers as Stabilizers for Conventional Polymers. Vol. 101, pp. 65–168.
Pospíšil, J.: Aromatic and Heterocyclic Amines in Polymer Stabilization. Vol. 124, pp. 87–190.
Powers, A. C. see Prokop, A.: Vol. 136, pp. 53–74.
Prasad, P. N. see Lin, T.-C.: Vol. 161, pp. 157–193.
Priddy, D. B.: Recent Advances in Styrene Polymerization. Vol. 111, pp. 67–114.
Priddy, D. B.: Thermal Discoloration Chemistry of Styrene-co-Acrylonitrile. Vol. 121, pp. 123–154.
Privalko, V. P. and *Novikov, V. V.*: Model Treatments of the Heat Conductivity of Heterogeneous Polymers. Vol. 119, pp. 31–78.
Prociak, A. see Bogdal, D.: Vol. 163, pp. 193–263.
Prokop, A., Hunkeler, D., DiMari, S., Haralson, M. A. and *Wang, T. G.*: Water Soluble Polymers for Immunoisolation I: Complex Coacervation and Cytotoxicity. Vol. 136, pp. 1–52.

Prokop, A., Hunkeler, D., Powers, A. C., Whitesell, R. R. and *Wang, T. G.*: Water Soluble Polymers for Immunoisolation II: Evaluation of Multicomponent Microencapsulation Systems. Vol. 136, pp. 53–74.
Prokop, A., Kozlov, E., Carlesso, G. and *Davidsen, J. M.*: Hydrogel-Based Colloidal Polymeric System for Protein and Drug Delivery: Physical and Chemical Characterization, Permeability Control and Applications. Vol. 160, pp. 119–174.
Pruitt, L. A.: The Effects of Radiation on the Structural and Mechanical Properties of Medical Polymers. Vol. 162, pp. 65–95.
Pudavar, H. E. see Lin, T.-C.: Vol. 161, pp. 157–193.
Pukánszky, B. and *Fekete, E.*: Adhesion and Surface Modification. Vol. 139, pp. 109–154.
Putnam, D. and *Kopecek, J.*: Polymer Conjugates with Anticancer Acitivity. Vol. 122, pp. 55–124.
Putra, E. G. R. see Ungar, G.: Vol. 180, pp. 45–87.

Quirk, R. P., Yoo, T., Lee, Y., M., Kim, J. and *Lee, B.*: Applications of 1,1-Diphenylethylene Chemistry in Anionic Synthesis of Polymers with Controlled Structures. Vol. 153, pp. 67–162.

Ramaraj, R. and *Kaneko, M.*: Metal Complex in Polymer Membrane as a Model for Photosynthetic Oxygen Evolving Center. Vol. 123, pp. 215–242.
Rangarajan, B. see Scranton, A. B.: Vol. 122, pp. 1–54.
Ranucci, E. see Söderqvist Lindblad, M.: Vol. 157, pp. 139–161.
Raphaël, E. see Léger, L.: Vol. 138, pp. 185–226.
Rastogi, S. and *Terry, A. E.*: Morphological implications of the interphase bridging crystalline and amorphous regions in semi-crystalline polymers. Vol. 180, pp. 161–194.
Ray, C. R. and *Moore, J. S.*: Supramolecular Organization of Foldable Phenylene Ethynylene Oligomers. Vol. 177, pp. 99–149.
Reddinger, J. L. and *Reynolds, J. R.*: Molecular Engineering of p-Conjugated Polymers. Vol. 145, pp. 57–122.
Reghunadhan Nair, C. P., Mathew, D. and *Ninan, K. N.*: Cyanate Ester Resins, Recent Developments. Vol. 155, pp. 1–99.
Reichert, K. H. see Hunkeler, D.: Vol. 112, pp. 115–134.
Rehahn, M., Mattice, W. L. and *Suter, U. W.*: Rotational Isomeric State Models in Macromolecular Systems. Vol. 131/132, pp. 1–475.
Rehahn, M. see Bohrisch, J.: Vol. 165, pp. 1–41.
Rehahn, M. see Holm, C.: Vol. 166, pp. 1–27.
Reineker, P. see Holm, C.: Vol. 166, pp. 67–111.
Reitberger, T. see Jacobson, K.: Vol. 169, pp. 151–176.
Reynolds, J. R. see Reddinger, J. L.: Vol. 145, pp. 57–122.
Richter, D. see Ewen, B.: Vol. 134, pp. 1–130.
Richter, D., Monkenbusch, M. and *Colmenero, J.*: Neutron Spin Echo in Polymer Systems. Vol. 174, pp. 1–221.
Riegler, S. see Trimmel, G.: Vol. 176, pp. 43–87.
Risse, W. see Grubbs, R.: Vol. 102, pp. 47–72.
Rivas, B. L. and *Geckeler, K. E.*: Synthesis and Metal Complexation of Poly(ethyleneimine) and Derivatives. Vol. 102, pp. 171–188.
Roberts, G. W. see Kennedy, K. A.: Vol. 175, pp. 329–346.
Robin, J. J.: The Use of Ozone in the Synthesis of New Polymers and the Modification of Polymers. Vol. 167, pp. 35–79.
Robin, J. J. see Boutevin, B.: Vol. 102, pp. 105–132.

Rodríguez-Pérez, M. A.: Crosslinked Polyolefin Foams: Production, Structure, Properties, and Applications. Vol. 184, pp. 97–126.
Roe, R.-J.: MD Simulation Study of Glass Transition and Short Time Dynamics in Polymer Liquids. Vol. 116, pp. 111–114.
Roovers, J. and *Comanita, B.*: Dendrimers and Dendrimer-Polymer Hybrids. Vol. 142, pp. 179–228.
Rothon, R. N.: Mineral Fillers in Thermoplastics: Filler Manufacture and Characterisation. Vol. 139, pp. 67–108.
de Rosa, C. see Auriemma, F.: Vol. 181, pp. 1–74.
Rozenberg, B. A. see Williams, R. J. J.: Vol. 128, pp. 95–156.
Rühe, J., Ballauff, M., Biesalski, M., Dziezok, P., Gröhn, F., Johannsmann, D., Houbenov, N., Hugenberg, N., Konradi, R., Minko, S., Motornov, M., Netz, R. R., Schmidt, M., Seidel, C., Stamm, M., Stephan, T., Usov, D. and *Zhang, H.*: Polyelectrolyte Brushes. Vol. 165, pp. 79–150.
Ruckenstein, E.: Concentrated Emulsion Polymerization. Vol. 127, pp. 1–58.
Ruiz-Taylor, L. see Mathieu, H. J.: Vol. 162, pp. 1–35.
Rusanov, A. L.: Novel Bis (Naphtalic Anhydrides) and Their Polyheteroarylenes with Improved Processability. Vol. 111, pp. 115–176.
Rusanov, A. L., Likhatchev, D., Kostoglodov, P. V., Müllen, K. and *Klapper, M.*: Proton-Exchanging Electrolyte Membranes Based on Aromatic Condensation Polymers. Vol. 179, pp. 83–134.
Russel, T. P. see Hedrick, J. L.: Vol. 141, pp. 1–44.
Russum, J. P. see Schork, F. J.: Vol. 175, pp. 129–255.
Rychly, J. see Lazár, M.: Vol. 102, pp. 189–222.
Ryner, M. see Stridsberg, K. M.: Vol. 157, pp. 27–51.
Ryzhov, V. A. see Bershtein, V. A.: Vol. 114, pp. 43–122.

Sabsai, O. Y. see Barshtein, G. R.: Vol. 101, pp. 1–28.
Saburov, V. V. see Zubov, V. P.: Vol. 104, pp. 135–176.
Saito, S., Konno, M. and *Inomata, H.*: Volume Phase Transition of N-Alkylacrylamide Gels. Vol. 109, pp. 207–232.
Samsonov, G. V. and *Kuznetsova, N. P.*: Crosslinked Polyelectrolytes in Biology. Vol. 104, pp. 1–50.
Santa Cruz, C. see Baltá-Calleja, F. J.: Vol. 108, pp. 1–48.
Santos, S. see Baschnagel, J.: Vol. 152, p. 41–156.
Sato, T. and *Teramoto, A.*: Concentrated Solutions of Liquid-Christalline Polymers. Vol. 126, pp. 85–162.
Schaller, C. see Bohrisch, J.: Vol. 165, pp. 1–41.
Schäfer, R. see Köhler, W.: Vol. 151, pp. 1–59.
Scherf, U. and *Müllen, K.*: The Synthesis of Ladder Polymers. Vol. 123, pp. 1–40.
Schlatmann, R. see Northolt, M. G.: Vol. 178, pp. 1–108.
Schmid, F., see Müller, M.: Vol. 185, pp. 1–58.
Schmidt, M. see Förster, S.: Vol. 120, pp. 51–134.
Schmidt, M. see Rühe, J.: Vol. 165, pp. 79–150.
Schmidt, M. see Volk, N.: Vol. 166, pp. 29–65.
Scholz, M.: Effects of Ion Radiation on Cells and Tissues. Vol. 162, pp. 97–158.
Schönherr, H. see Vancso, G. J.: Vol. 182, pp. 55–129.
Schopf, G. and *Koßmehl, G.*: Polythiophenes – Electrically Conductive Polymers. Vol. 129, pp. 1–145.

Schork, F. J., Luo, Y., Smulders, W., Russum, J. P., Butté, A. and *Fontenot, K.*: Miniemulsion Polymerization. Vol. 175, pp. 127–255.
Schulz, E. see *Munz, M.*: Vol. 164, pp. 97–210.
Schwahn, D.: Critical to Mean Field Crossover in Polymer Blends. Vol. 183, pp. 1–61.
Seppälä, J. see *Löfgren, B.*: Vol. 169, pp. 1–12.
Sturm, H. see *Munz, M.*: Vol. 164, pp. 87–210.
Schweizer, K. S.: Prism Theory of the Structure, Thermodynamics, and Phase Transitions of Polymer Liquids and Alloys. Vol. 116, pp. 319–378.
Scranton, A. B., Rangarajan, B. and *Klier, J.*: Biomedical Applications of Polyelectrolytes. Vol. 122, pp. 1–54.
Sefton, M. V. and *Stevenson, W. T. K.*: Microencapsulation of Live Animal Cells Using Polycrylates. Vol. 107, pp. 143–198.
Seidel, C. see *Holm, C.*: Vol. 166, pp. 67–111.
Seidel, C. see *Rühe, J.*: Vol. 165, pp. 79–150.
El Seoud, O. A. and *Heinze, T.*: Organic Esters of Cellulose: New Perspectives for Old Polymers. Vol. 186, pp. 103–149.
Shamanin, V. V.: Bases of the Axiomatic Theory of Addition Polymerization. Vol. 112, pp. 135–180.
Shcherbina, M. A. see *Ungar, G.*: Vol. 180, pp. 45–87.
Sheiko, S. S.: Imaging of Polymers Using Scanning Force Microscopy: From Superstructures to Individual Molecules. Vol. 151, pp. 61–174.
Sherrington, D. C. see *Cameron, N. R.*: Vol. 126, pp. 163–214.
Sherrington, D. C. see *Lin, J.*: Vol. 111, pp. 177–220.
Sherrington, D. C. see *Steinke, J.*: Vol. 123, pp. 81–126.
Shibayama, M. see *Tanaka, T.*: Vol. 109, pp. 1–62.
Shiga, T.: Deformation and Viscoelastic Behavior of Polymer Gels in Electric Fields. Vol. 134, pp. 131–164.
Shim, H.-K. and *Jin, J.*: Light-Emitting Characteristics of Conjugated Polymers. Vol. 158, pp. 191–241.
Shoda, S. see *Kobayashi, S.*: Vol. 121, pp. 1–30.
Siegel, R. A.: Hydrophobic Weak Polyelectrolyte Gels: Studies of Swelling Equilibria and Kinetics. Vol. 109, pp. 233–268.
de Silva, D. S. M. see *Ungar, G.*: Vol. 180, pp. 45–87.
Silvestre, F. see *Calmon-Decriaud, A.*: Vol. 207, pp. 207–226.
Sillion, B. see *Mison, P.*: Vol. 140, pp. 137–180.
Simon, F. see *Spange, S.*: Vol. 165, pp. 43–78.
Simon, G. P. see *Becker, O.*: Vol. 179, pp. 29–82.
Simonutti, R. see *Sozzani, P.*: Vol. 181, pp. 153–177.
Singh, R. P. see *Sivaram, S.*: Vol. 101, pp. 169–216.
Singh, R. P. see *Desai, S. M.*: Vol. 169, pp. 231–293.
Sinha Ray, S. see *Biswas, M.*: Vol. 155, pp. 167–221.
Sivaram, S. and *Singh, R. P.*: Degradation and Stabilization of Ethylene-Propylene Copolymers and Their Blends: A Critical Review. Vol. 101, pp. 169–216.
Slugovc, C. see *Trimmel, G.*: Vol. 176, pp. 43–87.
Smulders, W. see *Schork, F. J.*: Vol. 175, pp. 129–255.
Soares, J. B. P. see *Anantawaraskul, S.*: Vol. 182, pp. 1–54.
Sozzani, P., Bracco, S., Comotti, A. and *Simonutti, R.*: Motional Phase Disorder of Polymer Chains as Crystallized to Hexagonal Lattices. Vol. 181, pp. 153–177.
Söderqvist Lindblad, M., Liu, Y., Albertsson, A.-C., Ranucci, E. and *Karlsson, S.*: Polymer from Renewable Resources. Vol. 157, pp. 139–161.

Spange, S., Meyer, T., Voigt, I., Eschner, M., Estel, K., Pleul, D. and *Simon, F.*: Poly(Vinylformamide-co-Vinylamine)/Inorganic Oxid Hybrid Materials. Vol. 165, pp. 43–78.
Stamm, M. see Möhwald, H.: Vol. 165, pp. 151–175.
Stamm, M. see Rühe, J.: Vol. 165, pp. 79–150.
Starodybtzev, S. see Khokhlov, A.: Vol. 109, pp. 121–172.
Stegeman, G. I. see Canva, M.: Vol. 158, pp. 87–121.
Steinke, J., Sherrington, D. C. and *Dunkin, I. R.*: Imprinting of Synthetic Polymers Using Molecular Templates. Vol. 123, pp. 81–126.
Stelzer, F. see Trimmel, G.: Vol. 176, pp. 43–87.
Stenberg, B. see Jacobson, K.: Vol. 169, pp. 151–176.
Stenzenberger, H. D.: Addition Polyimides. Vol. 117, pp. 165–220.
Stephan, T. see Rühe, J.: Vol. 165, pp. 79–150.
Stevenson, W. T. K. see Sefton, M. V.: Vol. 107, pp. 143–198.
Stridsberg, K. M., Ryner, M. and *Albertsson, A.-C.*: Controlled Ring-Opening Polymerization: Polymers with Designed Macromoleculars Architecture. Vol. 157, pp. 27–51.
Sturm, H. see Munz, M.: Vol. 164, pp. 87–210.
Suematsu, K.: Recent Progress of Gel Theory: Ring, Excluded Volume, and Dimension. Vol. 156, pp. 136–214.
Sugimoto, H. and *Inoue, S.*: Polymerization by Metalloporphyrin and Related Complexes. Vol. 146, pp. 39–120.
Suginome, M. and *Ito, Y.*: Transition Metal-Mediated Polymerization of Isocyanides. Vol. 171, pp. 77–136.
Sumpter, B. G., Noid, D. W., Liang, G. L. and *Wunderlich, B.*: Atomistic Dynamics of Macromolecular Crystals. Vol. 116, pp. 27–72.
Sumpter, B. G. see Otaigbe, J. U.: Vol. 154, pp. 1–86.
Sun, H.-B. and *Kawata, S.*: Two-Photon Photopolymerization and 3D Lithographic Microfabrication. Vol. 170, pp. 169–273.
Suter, U. W. see Gusev, A. A.: Vol. 116, pp. 207–248.
Suter, U. W. see Leontidis, E.: Vol. 116, pp. 283–318.
Suter, U. W. see Rehahn, M.: Vol. 131/132, pp. 1–475.
Suter, U. W. see Baschnagel, J.: Vol. 152, pp. 41–156.
Suzuki, A.: Phase Transition in Gels of Sub-Millimeter Size Induced by Interaction with Stimuli. Vol. 110, pp. 199–240.
Suzuki, A. and *Hirasa, O.*: An Approach to Artifical Muscle by Polymer Gels due to Micro-Phase Separation. Vol. 110, pp. 241–262.
Suzuki, K. see Nomura, M.: Vol. 175, pp. 1–128.
Swiatkiewicz, J. see Lin, T.-C.: Vol. 161, pp. 157–193.

Tagawa, S.: Radiation Effects on Ion Beams on Polymers. Vol. 105, pp. 99–116.
Taguet, A., Ameduri, B. and *Boutevin, B.*: Crosslinking of Vinylidene Fluoride-Containing Fluoropolymers. Vol. 184, pp. 127–211.
Takata, T., Kihara, N. and *Furusho, Y.*: Polyrotaxanes and Polycatenanes: Recent Advances in Syntheses and Applications of Polymers Comprising of Interlocked Structures. Vol. 171, pp. 1–75.
Takeuchi, D. see Osakada, K.: Vol. 171, pp. 137–194.
Tan, K. L. see Kang, E. T.: Vol. 106, pp. 135–190.
Tanaka, H. and *Shibayama, M.*: Phase Transition and Related Phenomena of Polymer Gels. Vol. 109, pp. 1–62.
Tanaka, T. see Penelle, J.: Vol. 102, pp. 73–104.
Tauer, K. see Guyot, A.: Vol. 111, pp. 43–66.

Teramoto, A. see *Sato, T.*: Vol. 126, pp. 85–162.
Terent'eva, J. P. and *Fridman, M. L.*: Compositions Based on Aminoresins. Vol. 101, pp. 29–64.
Terry, A. E. see *Rastogi, S.*: Vol. 180, pp. 161–194.
Theodorou, D. N. see *Dodd, L. R.*: Vol. 116, pp. 249–282.
Thomson, R. C., Wake, M. C., Yaszemski, M. J. and *Mikos, A. G.*: Biodegradable Polymer Scaffolds to Regenerate Organs. Vol. 122, pp. 245–274.
Thünemann, A. F., Müller, M., Dautzenberg, H., Joanny, J.-F. and *Löwen, H.*: Polyelectrolyte complexes. Vol. 166, pp. 113–171.
Tieke, B. see v. *Klitzing, R.*: Vol. 165, pp. 177–210.
Tobita, H. see *Nomura, M.*: Vol. 175, pp. 1–128.
Tokita, M.: Friction Between Polymer Networks of Gels and Solvent. Vol. 110, pp. 27–48.
Traser, S. see *Bohrisch, J.*: Vol. 165, pp. 1–41.
Tries, V. see *Baschnagel, J.*: Vol. 152, p. 41–156.
Trimmel, G., Riegler, S., Fuchs, G., Slugovc, C. and *Stelzer, F.*: Liquid Crystalline Polymers by Metathesis Polymerization. Vol. 176, pp. 43–87.
Tsuruta, T.: Contemporary Topics in Polymeric Materials for Biomedical Applications. Vol. 126, pp. 1–52.

Uemura, T., Naka, K. and *Chujo, Y.*: Functional Macromolecules with Electron-Donating Dithiafulvene Unit. Vol. 167, pp. 81–106.
Ungar, G., Putra, E. G. R., de Silva, D. S. M., Shcherbina, M. A. and *Waddon, A. J.*: The Effect of Self-Poisoning on Crystal Morphology and Growth Rates. Vol. 180, pp. 45–87.
Usov, D. see *Rühe, J.*: Vol. 165, pp. 79–150.
Usuki, A., Hasegawa, N. and *Kato, M.*: Polymer-Clay Nanocomposites. Vol. 179, pp. 135–195.
Uyama, H. see *Kobayashi, S.*: Vol. 121, pp. 1–30.
Uyama, Y.: Surface Modification of Polymers by Grafting. Vol. 137, pp. 1–40.

Vancso, G. J., Hillborg, H. and *Schönherr, H.*: Chemical Composition of Polymer Surfaces Imaged by Atomic Force Microscopy and Complementary Approaches. Vol. 182, pp. 55–129.
Varma, I. K. see *Albertsson, A.-C.*: Vol. 157, pp. 99–138.
Vasilevskaya, V. see *Khokhlov, A.*: Vol. 109, pp. 121–172.
Vaskova, V. see *Hunkeler, D.*: Vol. 112, pp. 115–134.
Verdugo, P.: Polymer Gel Phase Transition in Condensation-Decondensation of Secretory Products. Vol. 110, pp. 145–156.
Vettegren, V. I. see *Bronnikov, S. V.*: Vol. 125, pp. 103–146.
Vilgis, T. A. see *Holm, C.*: Vol. 166, pp. 67–111.
Viovy, J.-L. and *Lesec, J.*: Separation of Macromolecules in Gels: Permeation Chromatography and Electrophoresis. Vol. 114, pp. 1–42.
Vlahos, C. see *Hadjichristidis, N.*: Vol. 142, pp. 71–128.
Voigt, I. see *Spange, S.*: Vol. 165, pp. 43–78.
Volk, N., Vollmer, D., Schmidt, M., Oppermann, W. and *Huber, K.*: Conformation and Phase Diagrams of Flexible Polyelectrolytes. Vol. 166, pp. 29–65.
Volksen, W.: Condensation Polyimides: Synthesis, Solution Behavior, and Imidization Characteristics. Vol. 117, pp. 111–164.
Volksen, W. see *Hedrick, J. L.*: Vol. 141, pp. 1–44.
Volksen, W. see *Hedrick, J. L.*: Vol. 147, pp. 61–112.
Vollmer, D. see *Volk, N.*: Vol. 166, pp. 29–65.
Voskerician, G. and *Weder, C.*: Electronic Properties of PAEs. Vol. 177, pp. 209–248.

Waddon, A. J. see Ungar, G.: Vol. 180, pp. 45–87.
Wagener, K. B. see Baughman, T. W.: Vol. 176, pp. 1–42.
Wake, M. C. see Thomson, R. C.: Vol. 122, pp. 245–274.
Wandrey, C., Hernández-Barajas, J. and *Hunkeler, D.*: Diallyldimethylammonium Chloride and its Polymers. Vol. 145, pp. 123–182.
Wang, K. L. see Cussler, E. L.: Vol. 110, pp. 67–80.
Wang, S.-Q.: Molecular Transitions and Dynamics at Polymer/Wall Interfaces: Origins of Flow Instabilities and Wall Slip. Vol. 138, pp. 227–276.
Wang, S.-Q. see Bhargava, R.: Vol. 163, pp. 137–191.
Wang, T. G. see Prokop, A.: Vol. 136, pp. 1–52; 53–74.
Wang, X. see Lin, T.-C.: Vol. 161, pp. 157–193.
Webster, O. W.: Group Transfer Polymerization: Mechanism and Comparison with Other Methods of Controlled Polymerization of Acrylic Monomers. Vol. 167, pp. 1–34.
Weder, C. see Voskerician, G.: Vol. 177, pp. 209–248.
Weis, J.-J. and *Levesque, D.*: Simple Dipolar Fluids as Generic Models for Soft Matter. Vol. 185, pp. 163–225.
Whitesell, R. R. see Prokop, A.: Vol. 136, pp. 53–74.
Williams, R. A. see Geil, P. H.: Vol. 180, pp. 89–159.
Williams, R. J. J., Rozenberg, B. A. and *Pascault, J.-P.*: Reaction Induced Phase Separation in Modified Thermosetting Polymers. Vol. 128, pp. 95–156.
Winkler, R. G. see Holm, C.: Vol. 166, pp. 67–111.
Winter, H. H. and *Mours, M.*: Rheology of Polymers Near Liquid-Solid Transitions. Vol. 134, pp. 165–234.
Wittmeyer, P. see Bohrisch, J.: Vol. 165, pp. 1–41.
Wood-Adams, P. M. see Anantawaraskul, S.: Vol. 182, pp. 1–54.
Wu, C.: Laser Light Scattering Characterization of Special Intractable Macromolecules in Solution. Vol. 137, pp. 103–134.
Wunderlich, B. see Sumpter, B. G.: Vol. 116, pp. 27–72.

Xiang, M. see Jiang, M.: Vol. 146, pp. 121–194.
Xie, T. Y. see Hunkeler, D.: Vol. 112, pp. 115–134.
Xu, P. see Geil, P. H.: Vol. 180, pp. 89–159.
Xu, Z., Hadjichristidis, N., Fetters, L. J. and *Mays, J. W.*: Structure/Chain-Flexibility Relationships of Polymers. Vol. 120, pp. 1–50.

Yagci, Y. and *Endo, T.*: N-Benzyl and N-Alkoxy Pyridium Salts as Thermal and Photochemical Initiators for Cationic Polymerization. Vol. 127, pp. 59–86.
Yamaguchi, I. see Yamamoto, T.: Vol. 177, pp. 181–208.
Yamamoto, T., Yamaguchi, I. and *Yasuda, T.*: PAEs with Heteroaromatic Rings. Vol. 177, pp. 181–208.
Yamaoka, H.: Polymer Materials for Fusion Reactors. Vol. 105, pp. 117–144.
Yannas, I. V.: Tissue Regeneration Templates Based on Collagen-Glycosaminoglycan Copolymers. Vol. 122, pp. 219–244.
Yang, J. see Geil, P. H.: Vol. 180, pp. 89–159.
Yang, J. S. see Jo, W. H.: Vol. 156, pp. 1–52.
Yasuda, H. and *Ihara, E.*: Rare Earth Metal-Initiated Living Polymerizations of Polar and Nonpolar Monomers. Vol. 133, pp. 53–102.
Yasuda, T. see Yamamoto, T.: Vol. 177, pp. 181–208.
Yaszemski, M. J. see Thomson, R. C.: Vol. 122, pp. 245–274.
Yoo, T. see Quirk, R. P.: Vol. 153, pp. 67–162.

Yoon, D. Y. see *Hedrick, J. L.*: Vol. 141, pp. 1–44.
Yoshida, H. and *Ichikawa, T.*: Electron Spin Studies of Free Radicals in Irradiated Polymers. Vol. 105, pp. 3–36.

Zhang, H. see *Rühe, J.*: Vol. 165, pp. 79–150.
Zhang, Y.: Synchrotron Radiation Direct Photo Etching of Polymers. Vol. 168, pp. 291–340.
Zheng, J. and *Swager, T. M.*: Poly(arylene ethynylene)s in Chemosensing and Biosensing. Vol. 177, pp. 151–177.
Zhou, H. see *Jiang, M.*: Vol. 146, pp. 121–194.
Zhou, Z. see *Abe, A.*: Vol. 181, pp. 121–152.
Zubov, V. P., Ivanov, A. E. and *Saburov, V. V.*: Polymer-Coated Adsorbents for the Separation of Biopolymers and Particles. Vol. 104, pp. 135–176.

Subject Index

Adhesive wear *II* 155
AFM, lamellae *II* 1
– spherulites *II* 1
Antiplasticizers *I* 39, 57, 106
– epoxy networks *I* 145
Aryl-aliphatic copolyamides *I* 111, 316
Aryl-aliphatic epoxy resins *I* 130
Asperities *II* 157

Ball/pin impression *II* 114
Bisphenol A *II* 5
– polycarbonate *I* 62, 296
BPA-PC *I* 62, 296
Branching *II* 1
– temperature *II* 19
Brittle-ductile transition *II* 227

Calcium pimelate *II* 46
Capillary pressure *II* 111
Carbonate motions *I* 85
Chain disentanglement craze (CDC) *I* 230
Chain scission *I* 1
Chain scission craze (CSC) *I* 230
Channel die compression *I* 12
Characteristic ratio *I* 4
Chloral-PC *I* 63
CMIMx *I* 185, 262
Cohesive wear *II* 155
Compact tension *I* 241, *II* 120
Conformation *I* 224
Contact fatigue cracking *II* 166
Copolyamides, aryl-aliphatic *I* 111, 316
Crack, pre-crack *I* 242
Crack acceleration *I* 16
Crack growth, slow *I* 22, 28
– steady *I* 15
Crack growth/propagation *II* 97, 178
– directions *II* 174
Crack initiation *II* 171

Crack opening displacement (COD) *II* 121, 226
Crack propagation *I* 1, 240
Crack-tip stress *II* 120
Craze breakdown *II* 207, 214
Craze fibril breakdown *I* 231
Craze initiation *II* 204, 213
Craze morphology *I* 228
Craze nucleation *I* 22
Craze thickening *II* 205, 214
Crazes *I* 1
– intrinsic *I* 25
– morphology *I* 22
Crazing *II* 109, 197, 203
– cohesive surface model *II* 212
Creep *I* 1, 16
Creep curves, POM *I* 27
Creep deformation *I* 17, 20, 27
– semicrystalline polymers *I* 27
Creep failure *I* 21
Creep strain *I* 17
Critical displacement amplitude *II* 164
Critical strain test *II* 115
Critical stress intensity factor *II* 116
Crystallinity, ESC *II* 133
Crystallization *I* 7
Cyclohexylmaleimide (CMI) *I* 178, 262

Damage mechanics *I* 30
Deformation *I* 3, 18
DGEBA *I* 131, *II* 166, 183
N,N-Dicyclohexyl-2,6-naphthalene dicarboxamide *II* 46
Dimethyl terephthlate (DMT) *I* 58
Diphenyl carbonate *I* 91, 92
Diphenyl propane *I* 91, 92
Disentanglement *I* 2
– rate *I* 29
Displacement amplitude *II* 164

Ductile failure *I* 13
Ductile-brittle transitions *II* 178, 187
Dugdale model *II* 208
Dynamic mechanical analysis (DMA)
 II 85

Embrittlement *II* 108
Embryo *II* 1
Energy to break *II* 65
Entanglements *I* 6, 29, 243
– density *I* 8
Environmental stress cracking *II* 108
Epoxies, siloxane-modified *II* 183
Epoxy networks, antiplasticizers *I* 145
Epoxy networks, toughened *II* 182
– contact fatigue *II* 163
Epoxy resins, aryl-aliphatic *I* 130
ESCR *II* 108, 113
Essential work of fracture *II* 60
Ethylene, ESR *II* 114
Eyring equation *I* 223
Eyring's reaction rate theory *I* 17

Failure *I* 1, 21
– creep *I* 21
– ductile *I* 13
Failure strength, time-dependent *II* 110
Fatigue crack growth *II* 105, 116
– initiation *II* 119
– propagation *II* 116
Fatigue resistance *II* 105
Fibril breakdown *I* 231
Fibrillation energy *I* 26
Films, microporous, PP *II* 95
Flow stress *I* 249
Founding lamella *II* 9
Fracture, dynamic/elastic *I* 14
– PMMA *I* 258
Fracture mechanics, linear elastic *I* 13
Fretting *II* 164
– loading, contact conditions *II* 166
FTIR *I* 224

Glass-rubber transition *I* 41
Glutarimide *I* 156, 190, 244, 272, *II* 186
Gross slip condition *II* 163

HA60/HA95 *I* 138
Hardening *I* 18
Hertzian cracks *II* 179

Hexamethylene diamine *I* 131
Hildebrand solubility parameter *II* 112
HIPS *II* 136
Hydrostatic pressure *I* 87
Hydroxypropyl ether *I* 141

Indenter geometries *II* 159
Induction period *I* 13
Inertia peak *I* 15
Isophorone diamine (IPD) *II* 166
Isopropyl alcohol (IPA) *II* 129

Jumps *I* 17

Lamellae *II* 1
– branching *II* 14
– founding *II* 9
– growth rate *II* 25
– film thickness *II* 29
– β-PP *II* 89
– propagation *II* 22
Linear elastic fracture mechanics *I* 13

Main-chain reorientation *I* 85
Maleimide *I* 156, 244
Melt flow rate *II* 51
Methyl glutarimide *I* 190
Methyl methacrylate-co-*N*-
 cyclohexylmaleimide *I* 178,
 262
Microvoids, β-PP *II* 83
Mindlin's analysis *II* 164
MMA *I* 156, 262
MMA, fatigue *II* 185
MMA-glutarimide *II* 186
MMA-maleimide *II* 186
Mobility *I* 3
Molecular dynamics *I* 101
Molecular mobility *I* 3
Motional processes *I* 43
MT copolyamide *I* 119
Multiaxial stress loading *II* 141
MWD *II* 54

Notched impact strength *II* 55
Nucleating agent *II* 57
Nucleation *I* 22
Nucleation, heterogeneous *II* 37
– homogeneous *II* 6

Subject Index

Partial slip condition II 163
PE II 108
PE-HD II 132
PE-LD II 133
PEO II 5
PET I 51
Phenyl ring flips I 96
Pipes, PP II 96
Plastic event I 7
Plastic flow I 227, 265, 273
Plastic flow stress, PMMA I 249
PMMA I 156, 217, 244, II 128, 138, 157, 187, 204
– fracture I 258
– plastic deformation I 245
– plastic flow stress I 249
– stress-strain curves I 246
– thin film I 256
– Wöhler diagram II 119
PMMA-kerosene II 113
Poly(cycloalkyl methacrylates) I 45
Poly(ethylene terephthalates) I 51
Poly(ethylene vinyl acetate) II 33
Poly(methyl methacrylate) I 156
Poly(phenylene oxide) II 188
Polycaprolactone II 7, 8, 14
Polycarbonate I 62, 296
Polycarbonate, scratching map II 160
Polydispersity II 54
Polypropylene, filled, toughness II 70
– β-nucleated II 43
– rubber-toughed II 73
– toughness II 50, 51
Polysiloxane II 183
Polystyrene, atactic I 225, 229
Polystyrene, brittle failure modes II 161, 188
– stresscracking II 127
POM, creep curves I 27
PP II 14
– glass-fiber reinforced II 94
PP fibers II 93
PP/EPR II 76
Precrack I 242, II 120
PTFE II 155
PVC II 108, 130

Quinacridone II 46

Random coil I 4
Rapid crack propagation II 97

Reaction rate theory, Eyring I 17
Reorientation, main-chain I 85
Rotational isomeric state I 94
Rubber particles, stress concentrators II 138

Scratch tests II 158
Secondary transitions I 39
Semicrystalline polymers, creep deformation I 27
– slow crack growth I 28
Shear deformation zone (SDZ) I 228, 232, 269
Shear yielding II 197
Sherby-Dorn plot I 18
Siloxane elastomers II 182
Sliding contacts, fracture II 153
Slow crack growth II 97
Softening I 227, 251, 265, 275
Solid-state transitions I 41
Spherulites II 1
– formation II 35
Steady crack growth I 15
Stiffness, C–C bond I 4
Strain rate, local I 22
Strain softening I 227, 251, 265, 275
Stress cracking II 105
Stress intensity factor I 14
Stress transfer, lamellae I 11
Stress transmission, breakdown I 22
Stress-state II 64
Stress-strain curves I 222
– PMMA I 246
Styrene-butadiene II 136
Suberate II 46
Surface energy I 24

Tensile stress II 114
Terephthlates I 51, 58
Tetrachlorophthalic dimethyl ester I 58
Tetramethyl bisphenol A polycarbonate I 62, 296
Thermoforming, PP II 94
Thin film I 268, 280
Thin layer yield I 12
Three-point bending I 242
Tie molecules I 7, II 132
TMBPA-PC I 62, 296, 306
Toughness I 1
TPDE I 58

Traction *I* 30
Transfer wear *II* 155
Transitions, secondary *I* 39
– solid-state *I* 41

UHMWPE *II* 155
– sterilization *II* 146

Viscoplastic deformation, amorphous polymers *II* 198

Viscosity, Eyring *I* 223

Weakening *I* 18
WLF expression *I* 41
Wöhler $S - N$ curves *II* 118

Yield point *I* 17
Yield stress *I* 223, 247, 263, 273, 299, *II* 87
Young modulus *II* 88

Printing: Krips bv, Meppel
Binding: Stürtz, Würzburg

DATE DUE

SCI QD 281 .P6 F6 no.188

Intrinsic molecular
mobility
 and toughness...